AQA KS3 Science 2 Extend PRACTICE BOOK

Cliff Curtis
Deborah Lowe
John Mynett

HODDER EDUCATION
AN HACHETTE UK COMPANY

Hachette UK's policy is to use papers that are natural, renewable and recyclable products and made from wood grown in well-managed forests and other controlled sources. The logging and manufacturing processes are expected to conform to the environmental regulations of the country of origin.

Orders: please contact Hachette UK Distribution, Hely Hutchinson Centre, Milton Road, Didcot, Oxfordshire, OX11 7HH. Telephone: +44 (0)1235 827827. Email education@hachette.co.uk. Lines are open from 9 a.m. to 5 p.m., Monday to Friday. You can also order through our website: www.hoddereducation.co.uk

First published in 2018 by
Hodder Education,
An Hachette UK Company
Carmelite House
50 Victoria Embankment
London EC4Y 0DZ

www.hoddereducation.co.uk

Impression number 10 9 8 7 6 5 4 3

Year 2023

Cover photo © PCN Photography / Alamy Stock Photo

Typeset in 11/14 pt Vectora LT Std 45 Light by Integra Software Services Pvt. Ltd., Pondicherry, India

Printed and bound by CPI Group (UK) Ltd, Croydon, CR0 4YY

A catalogue record for this title is available from the British Library.

ISBN: 9781510402515

Contents

Find the answers at www.hoddereducation.co.uk/AQAKS3Science

1 Contact forces

» Combining forces

Worked example

The car below is initially stationary. The following forces are then applied.

50 N   150 N

a) Calculate the resultant force on the object.
b) Is the car in equilibrium?
c) What will happen to the motion of this object?

a) As the forces are in opposite directions, the resultant force is equal to the largest force minus the smallest force. Therefore, resultant force = 150 − 50 = 100 N to the right.
b) The car is not in equilibrium, because the resultant force is not zero.
c) The object will start to move to the right.

Know

1 What does the term 'resultant force' mean?

2 What is the resultant force of an object in equilibrium?

3 What will happen to an object in equilibrium if it is:

   a) stationary

   b) travelling at a steady speed?

Apply

1 Look at the diagrams below.

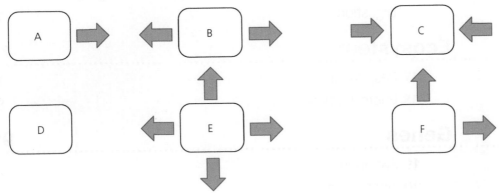

a) Which of the objects are in equilibrium?

b) If all of the objects were stationary to begin with, what would happen to the motion of each object when these forces were applied?

c) If all of the objects were travelling to the right at a steady speed to begin with, what would happen to the motion of each of object when these forces were applied?

2 Calculate the resultant force on each of the objects below.

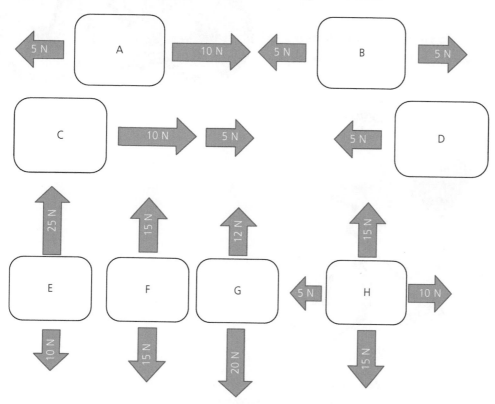

Extend

1 A car is travelling eastwards at a steady speed.

a) If the engine force of the car is 200 N, what must the frictional forces of the car (acting in the opposite direction) be?

b) If the frictional forces increase to 250 N, what is the resultant force on the car?

c) What would happen to the motion of the car now?

d) When the car finally stops, what must the size of the engine and frictional forces be? Why?

FORCES

» Force diagrams

Worked example

Draw the forces acting on a bike as it travels at a steady speed. The weight of the bike (and cyclist) is 1000N. The frictional forces (from the road and drag) working against the cyclist equal 500N.

The question tells us about two forces. Weight acts downwards, and is twice as large as the frictional forces, which are against the cyclist.

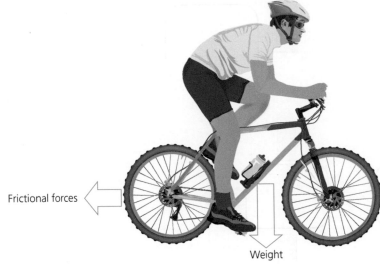

Frictional forces

Weight

If an object is travelling at a steady speed, it must be in equilibrium. Therefore, there must be a force that is an equal size to weight but acting upwards, and a force that is equal to the frictional forces but acting forwards.

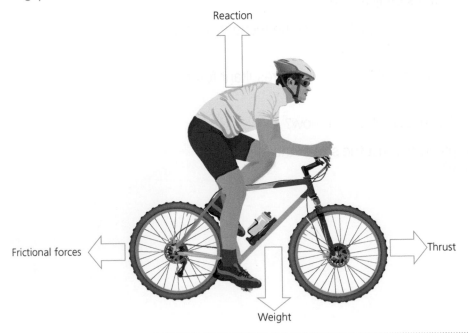

Reaction

Frictional forces

Thrust

Weight

Know

1 What three things does a force arrow tell us about a force, and how?

2 Add in the forces that must be present for the following objects to be in equilibrium.

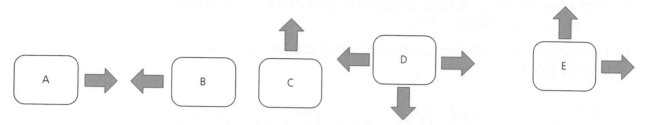

Apply

1 Draw and label the forces on the following objects.

   a) A book resting on a table.

   b) A car travelling at a steady speed.

   c) A skydiver speeding up as she falls to Earth.

   d) A rubber duck floating in a bath.

   e) A car decelerating (slowing down).

2 Draw force diagrams for the following situations, calculate the resultant force and describe what would happen to the motion of each object.

   a) A car with an engine force forwards of 300N and a drag force backwards of 500N.

   b) An anchor on the seabed with a weight of 200N and a reaction force that balances this.

   c) A rocket with a thrust force of 1000N and a weight of 800N.

   d) A boat with a weight of 10000N and an upthrust force of 10000N.

Extend

1 Aeroplanes mainly experience five forces – thrust (from the engine), lift (from the wings), weight (due to gravity), drag (from air resistance, which opposes its motion) and the normal contact force (if it is on the ground). Draw and explain the forces on an aeroplane as it:

   a) rests on the runway, before the engine is turned on

   b) takes off

   c) flies at a steady speed and height

   d) descends for landing.

2 A skydiver jumps out of a plane. Draw force diagrams for the following situations and explain how they will affect the skydiver's motion.

a) As the skydiver jumps out of the plane, the only force acting on him is his weight due to gravity. This is 500 N.

b) After a few seconds, the skydiver opens his parachute. This eventually causes him to be in equilibrium.

c) The skydiver eventually hits the ground, which provides a reaction force of 1000 N.

» Effects of forces on shape

Worked example

A spring is originally 15 cm long. When two 100 g masses are added to the end of it, its length increases to 25 cm.

a) What is the force applied to the spring in this situation?
b) What is the extension of the spring?
c) How long would the spring be if a third 100 g mass was added?

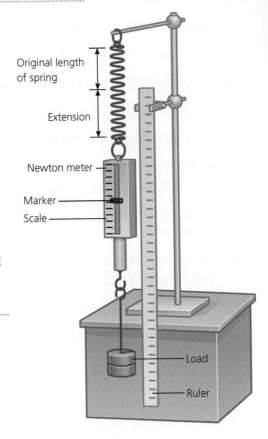

Original length of spring

Extension

Newton meter

Marker

Scale

Load

Ruler

a) The force is caused by the weight of the masses:

weight = mass × gravitational field strength = 0.2 kg × 10 = 2 N

b) Extension is how much the length has increased by:

extension = 25 cm – 15 cm = 10 cm

c) Two masses cause a 10 cm extension, so one mass would cause half of this, 5 cm. Therefore the spring would have a length of 25 cm + 5 cm = 30 cm.

Know

1 When does deformation occur?

2 What is the difference between tension and compression?

3 What is the difference between the independent and the dependent variable in an investigation?

4 Give some examples of when forces can deform objects.

Apply

1 When a force (or load) is applied to a spring, it stretches. If we increase that load, the extension of the spring increases too. We can sketch a graph of this experiment as shown on the right.

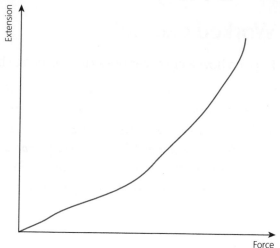

a) Which is the independent variable in the experiment and which is the dependent variable? Why?

b) How does the graph show that the force applied is directly proportional to the extension (at least to start with)?

c) Where is the elastic limit of the spring shown on the graph?

d) What will happen to the spring if it is stretched past its elastic limit?

e) How would the graph look different for a spring with a higher elastic limit?

f) How would the graph look different for a more stretchy spring?

Extend

1 The following data were collected when stretching two different springs by applying a force to them.

Spring A Applied force (N)	Extension (cm)			
	1	2	3	Average
2 N	1.9	2.1	2.0	
4 N	4.0	4.0	4.2	
6 N	5.9	5.8	6.0	
10 N	9.9	10.2	9.9	
12 N	14.0	14.3	14.5	
14 N	18.1	17.9	18.1	

Spring B Applied force (N)	Extension (cm)			
	1	2	3	Average
1 N	1.5	1.4	1.6	
3 N	4.4	4.4	4.4	
6 N	9.2	8.8	9.0	
8 N	11.9	11.9	12.2	
11 N	16.3	16.5	16.6	

a) Calculate the average extension for each force for both springs.

b) Draw an extension–force graph for each of the springs, on the same graph, and draw a line of best fit for each set of data.

c) What would the extension be for each spring when 5 N of force is applied?

d) What force is required to stretch each spring by 5 cm?

e) If spring A was 10 cm long and spring B was 20 cm long to start with, how long would they each be when 9 N of force is applied?

f) Which spring has an elastic limit? Where?

g) Which spring is the stiffest? How do you know?

» Drag

Worked example

Explain how a skydiver uses drag forces to help them reach the ground safely.

When a skydiver jumps out of an aeroplane, their weight causes them to accelerate. The drag force working against them is air resistance, which is caused by the air molecules rubbing past them. To help them slow down and not hit the ground too quickly, the skydiver will open a parachute. This increases their surface area, and so increases the drag force on them so that it is bigger than the downward force of weight. This slows them down, so they can hit the ground at a safe speed.

Know

1 Which of the following is not a fluid: solid, liquid, gas? Why?

2 When do drag forces occur?

Apply

1 Give two examples of when drag forces are useful, and two examples of when they are not.

2 Draw force diagrams to show how the drag forces on a lorry would be different from those on a racing car.

Extend

1 Cyclists focus on reducing drag. Using research if necessary, explain how the following help reduce the drag on a cyclist.

   a) curved helmets

   b) skin-tight body suits

   c) cycling in a pack or long line

» Levers

Worked example

Explain why it is easier to tighten a bolt with a spanner rather than just your hand.

A spanner is an example of a lever, which is something that allows the force that you apply to a situation to have a greater effect. When you try to tighten the bolt with your hand, you have to put a lot of force onto it to make it turn. When you turn it with a spanner, the force you apply is multiplied by the length of the spanner, giving a greater effect. The longer the spanner, the easier the bolt will be to turn.

Know

1 What do levers allow small forces to do?

2 Give some examples of levers in everyday life.

Apply

1 State two ways in which you can increase the turning effect of a force.

2 Person A and person B sit on opposite ends of a see-saw. Person A sits 3m away from the pivot in the centre. If person B is twice as heavy as person A, how far away from the pivot must person B be for the see-saw to be balanced?

Extend

1 The following children sit on a see-saw.

Child	Weight (N)	Distance from pivot (m)
A	500	2
B	400	2.5
C	250	4
D	200	5
E	300	1

a) Which child do you think will provide the smallest turning effect? Why?

b) Would the see-saw balance if child A sat on one side and child C sat on the other? If not, what would happen?

c) Identify two pairs of children who could sit on opposite ends of the see-saw and balance it.

d) What would happen if both child C and child D sat on one side of the see-saw and child A sat on the other?

e) Predict the weight of a child sitting 2m away from the pivot who could balance child E.

2 Pressure

» Floating and sinking

Worked example

A small rowing boat displaces 800 N of water when it is put into a lake.

a) What must the weight of the rowing boat be, if it is floating?
b) What is the mass of the boat?

a) If the boat is floating, the weight of water displaced must equal the weight of the boat. Therefore, the boat must weigh 800 N.

b) The mass of the boat can be calculated using the equation:

weight = mass × gravitational field strength

Therefore, rearranging the equation in terms of mass:

$$mass = \frac{weight}{gravitational\ field\ strength}$$

$$mass = \frac{800\,N}{10\,N/kg} = 80\,kg$$

Know

1 What is upthrust?

2 What do we mean when we say that water is displaced when an object is put in it?

3 Copy and complete the diagrams below by adding in an arrow showing the direction and size of the upthrust force for each situation.

Note: The anchor is sinking.

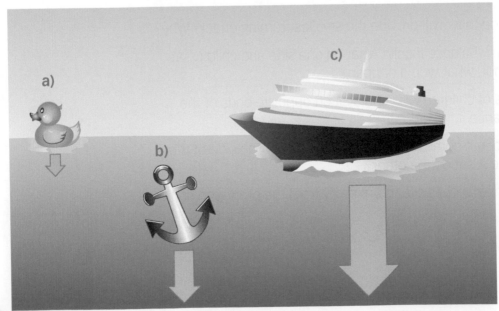

4 What equation relates mass and weight?

5 Copy and complete the sentences below.

  a) We can work out whether an object will float or sink by comparing the _____ of the object with the _____ of the water it _____.

  b) If the weight of the object is equal to weight of water it displaces, it will _____.

  c) If the weight of water is less than the weight of the object, it will _____.

Apply

1 Copy and complete the table below. The first row has been completed for you.

Mass of object	Weight of object	Weight of water the object must displace if it is to float
1 kg	10 N	10 N
500 g	5 N	
0.2 kg		
15 kg		150 N
	200 N	
		8 N
300 g		
	0.5 N	

2 For each of the objects below, draw a force diagram for the situation and decide whether the object will float or sink. Then calculate the mass of each object.

  a) A cork weighs 0.1 N and displaces 0.1 N of water.

  b) A pineapple weighs 5 N and displaces 3 N of water.

  c) A football weighs 7 N and displaces 7 N of water.

  d) A bar of soap weighs 2 N and displaces half of that weight of water.

3 A 1000 kg boat must displace what weight of water if it is to float?

4 A lily pad has a mass of 5 kg and displaces 50 N of water. Will it float? Why?

5 A sponge has a mass of 200 g and displaces 1.5 N of water. Will it float? Why?

Extend

1 An unpeeled orange floats, but a peeled one sinks (try this out for yourself if you don't believe it!). What can be understood about the amount of water that must be displaced in these two situations?

2 When you push a ball down into the water and let go, it rushes back up again. Why do you think this is, in terms of the forces acting on the ball? Draw a force diagram to help you explain.

3 Two teams of students make a raft in a team-building exercise. Both rafts have a mass of 50 kg, but only team A's raft floats.

a) What weight of water must raft A displace?

b) Would team B's raft displace more or less water than this?

c) Draw force diagrams to compare the forces acting on the two rafts.

d) Predict and discuss what could be different about team A's raft compared with team B's that helps it to float.

4 Boats often sink if they spring a leak. Try to explain why you think this could be, using ideas of displaced water weight.

» Forces on a surface

Worked example

A hammer hits a nail with a force of 300 N. The head of the nail has a surface area of 0.0005 m², whilst the sharp end has a surface area of 0.00002 m². Calculate the stress exerted on the wall.

For the end of the nail:

$$stress = \frac{force}{area} = \frac{300\,N}{0.00002\,m^2} = 15\,000\,000\,N/m^2$$

> **Hint**
>
> The area must always be in m² for this equation to work.

Know

1 What is the area of the following situations, in m²?

a) A triangle of base 4 m and height 2 m.

b) A square with sides that are each 50 cm long.

c) A rectangle with sides that are 35 mm and 15 mm long.

2 State two examples of high-stress situations and two examples of low-stress situations.

> **Hint**
>
> It is much easier to convert into metres before calculating the area, than converting into m² afterwards.

Apply

1 A drawing pin has a head that is approximately 0.005 m² and an end that is approximately 0.0002 m². You push it into a wall with a force of 50 N. What stress does the pin exert on:

a) your thumb

b) the wall?

2 Four children are standing on the floor, see the table on the right.

a) Predict and explain who you think will exert the most stress on the floor.

b) Calculate the stress that each child exerts on the floor. Was your prediction correct?

Child	Weight (N)	Area of feet (m²)
Sarah	400	0.022
Nino	500	0.036
Jim	450	0.027
Aalia	300	0.016

3 You are standing on a frozen pond and the ice is about to crack due to the stress you are exerting on it. How could you reduce the chance of falling into the water? Explain why.

Extend

1 A rhino has a mass of 2000 kg. Each of its feet has an area of $0.12\,m^2$.

    a) Calculate the stress the rhino must exert on the floor.

    b) Explain why the rhino sinks when it is standing on mud but not when it is standing on dry ground.

2 People usually find wearing high-heeled shoes more uncomfortable than wearing flat ones. Explain why, using ideas of stress.

3 Estimate the stress that you exert on the floor when you are:

    a) standing on one leg

    b) doing a handstand

    c) lying down like a starfish.

» Pressure in a fluid

Worked example

A diver has a surface area of $0.4\,m^2$. When they are 2 m below sea level, there is 800 kg of water above them. What pressure do they experience due to this?

For this, we need to use the equation:

$$\text{pressure} = \frac{\text{force}}{\text{area}}$$

The force here is the weight of the water above the diver. We calculate this using the equation:

$$\text{weight} = \text{mass} \times g = 800 \times 10 = 8000\,N$$

Then:

$$\text{pressure} = \frac{\text{force}}{\text{area}} = 8000/0.4 = 20\,000\,Pa$$

Know

1 What is pressure?

2 What is the equation for pressure?

3 What three units can be used for pressure?

4 How many pascals are in a kilopascal?

5 What do we mean by 'atmospheric pressure'?

6 How many pascals are in 1 atmosphere of pressure?

7 Put these in order of how much pressure you would feel from the environment if you were there, from highest to lowest.

   A at sea level

   B at the top of Mount Everest

   C at the bottom of the Atlantic Ocean

   D at the top of the Empire State Building

   E at the bottom of a swimming pool

Apply

1 Compare the similarities and differences between stress and pressure.

2 Describe how fluids can cause pressure.

3 The area of your head and shoulders is approximately $0.1\,m^2$. If atmospheric pressure is 100 kPa at sea level, what is the weight of the air molecules above you?

4 Water pressure increases by about 1 atmosphere for every 10 m that you travel beneath the water's surface. What would the water pressure on you be, in atmospheres and kPa at:

   a) 10 m below the surface

   b) 25 m below the surface

   c) 100 m below the surface (a recommended limit for divers)?

5 The area of one side of a person is about $0.8\,m^2$. Knowing this, calculate the weight of water pushing down on a diver for each of the water depths in question 4.

6 Explain why air pressure is lower at the top of a mountain compared with at sea level.

7 Hydraulic systems require an incompressible fluid to work. Use ideas about particle theory to explain why this fluid must be a liquid rather than a gas.

Extend

1 Which would exert a higher pressure – a room full of a gas or a room full of a liquid? Explain why.

2 When you are in an aeroplane, the cabin has to be pressurised as you fly, so you can breathe and act normally. Explain why this is necessary.

3 Submarines are specially designed to withstand the high pressures that come with travelling deep under water. Research the features that they have to help them do this.

4 If you make three small holes at different heights in a plastic bottle, and then fill it up with water, this happens (see the diagram on the right).

   Use ideas of water pressure and hydraulics to help you explain why this is the case.

5 Research how hydraulic brakes in a car work, and the advantages of using them over mechanical brakes.

3 Electromagnets

» Stronger electromagnets

Worked example

Some students make two electromagnets.

- Electromagnet A is made of 10 coils of wire and is connected to two 1.5V cells.
- Electromagnet B is made of 20 coils of wire and is connected to one 1.5V cell.

Which electromagnet would have the strongest magnetic field surrounding it?

This is a bit of a trick question! The strength of an electromagnet relies mainly on two factors – the number of coils in it and the size of the current passing through it, which depends on the voltage supplied to the circuit. Electromagnet A has half the number of coils of electromagnet B so would have a smaller magnetic field around it if they were both carrying the same current. However, electromagnet A is connected to twice the number of cells, and so will have twice the current passing through. This, in effect, 'cancels out' the effect of only having half the number of coils. Therefore, electromagnets A and B would both have the same strength magnetic field around them.

Know

1 What is electromagnetism?

2 Do electromagnets follow the same rules as permanent magnets when it comes to attraction and repulsion?

3 'Only iron, nickel and cobalt can be made into an electromagnet.' Is this statement correct?

4 When are electromagnets used instead of permanent magnets?

5 How can you increase or decrease the current in an electric circuit?

6 What dangers do the following pose?

   a) high voltage

   b) high current

7 What is a solenoid? Draw a diagram to help you explain.

8 State three ways to increase the strength of an electromagnet.

Apply

1 Compare an electromagnet with a permanent magnet.

2 Wrapping a wire around an iron nail can make a simple electromagnet (see the diagram on the right).

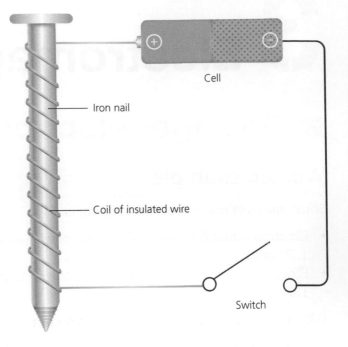

Iron nail

Cell

Coil of insulated wire

Switch

What would happen to the magnetic field around the electromagnet in the circuit shown if:

a) more coils were added to the wire

b) the cell was turned around

c) another cell was added

d) some coils were taken away

e) the switch was opened?

3 Explain why a solenoid causes a stronger magnetic field than a piece of straight wire.

4 Sketch the fields around the following solenoids, assuming they carry the same size current in each case.

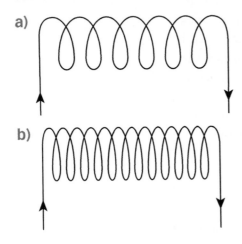

a)

b)

Extend

1 Research how electromagnets are used in:

a) electric bells

b) recycling plants

c) loudspeakers.

2 Explain why the wires used to make solenoids must be insulated.

3 Predict what will happen when two identical solenoids are placed next to each other when a current runs through each of them:

a) in the same direction

b) in opposite directions.

» Using electromagnets

Worked example

A student needs to separate a mixture of sand and iron filings into two piles. Explain the advantages of using an electromagnet to do this over a permanent magnet.

The student can use any type of magnet to separate the mixture – the magnetic iron filings will be attracted to the magnet, leaving the non-magnetic sand behind. The advantage comes in the second step – it will be very hard to take the iron filings off the permanent magnet, as they will always be attracted to it. Using an electromagnet would be better here because when the power supply is switched off the electromagnet will stop being magnetic, causing the iron filings to fall off the magnet into a separate pile.

Know

1 State two advantages of electromagnets over permanent magnets.

2 Electromagnets are used in recycling plants to sort out rubbish. Which of the following objects would the electromagnet pick up?

A plastic cups   D steel cans   G wooden table

B tinfoil trays   E aluminium cans

C newspapers   F old fridge

3 State three other uses of electromagnets in everyday life.

Apply

1 Loudspeakers use permanent and electromagnets to work. Put the following sentences in the correct order to explain how.

A The cone (containing the electromagnet) moves towards the permanent magnet.

B This causes a magnetic field around the solenoid.

C This causes a force between the electromagnet and the permanent magnet.

D A current runs through the electromagnet.

2 Explain how the force between the electromagnet and the permanent magnet in a loud speaker can be:

a) increased and decreased in strength

b) switched between attraction and repulsion.

Extend

1 Describe the function of relay switches and explain why they are used.

2 Research the use of relay switches in everyday life.

3 Research how nuclear power stations rely on electromagnets for safety during a power cut or natural disaster.

4 Magnetism

» Magnetism and navigation

Worked example

The north pole of a magnet is held near three sealed boxes containing hidden objects – another magnet, an iron nail and a wooden block. A student experiments with the boxes and finds that:

- box A is attracted to the magnet
- box B is not attracted to the magnet
- box C is repulsed from the magnet.

Which object is in which box?

Magnets can attract either other magnets or magnetic materials, so box A must contain either the other magnet or the iron nail (as iron is a magnetic material). However, we cannot yet tell which one it is. Box B is unaffected by the magnet, so must contain the non-magnetic material – the wooden block.

The magnet repels box C. Only another magnet can be repelled by a magnet, so box C must contain the second magnet. That means that box A must contain the iron nail.

Know

1 Which three metals are magnetic?

2 What is a magnetic field?

3 Why is magnetism an example of a non-contact force?

4 What is the difference between an attractive force and a repulsive force?

Apply

1 What would happen in the following situations? Will they attract, repulse or do nothing?

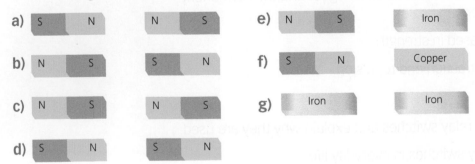

2 Below is a pair of horseshoe magnets.

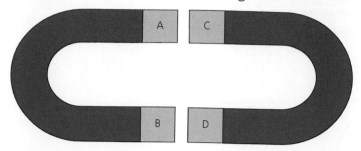

What must the poles A–D be if:

a) A is a north pole and the two magnets are attracted to each other

b) B is a north pole and the two magnets are repelled from each other?

3 Screwdrivers often have a small magnet at their tip. Explain why this might be useful.

> **Hint**
> ....
> Think about what most screws are made of.

Extend

1 Explain why the north pole of a compass needle does not actually point to the magnetic north pole of the Earth.

2 Research how two different animals use magnetism to help them navigate.

3 In Book 1, Chapter 4 you learned about electrostatics and charges. Compare the electrostatic force with the magnetic force.

» Magnetic fields

Worked example

Below is a diagram of the Earth's magnetic field.

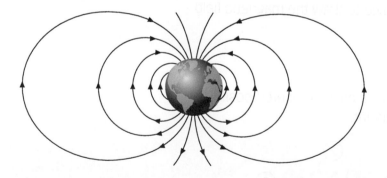

a) Label the magnetic north and south poles.
b) Indicate where on the Earth its magnetic field is weakest and explain your decision.

a) The arrows of the magnetic field lines always point from the north pole to the south pole. Therefore, the magnetic south pole is at the top of the Earth (the geographic north pole) and the magnetic north pole is at the bottom (the geographic south pole).

b) The strength of the field is shown by how close or far apart the field lines are. The lines are closes together at the poles and furthest apart at the sides, over the Equator. Therefore, the field is weakest at the Equator.

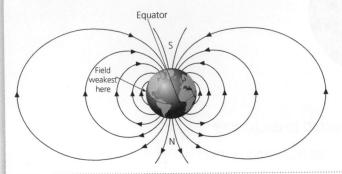

Know

1 Which materials can block a magnetic field?

2 Describe what is being shown in the diagram below.

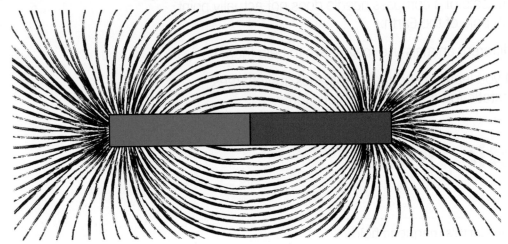

3 Describe how a plotting compass can be used to draw the magnetic field of a magnet.

Apply

1 Copy the diagrams below and add in arrows at each numbered position to show which way the compasses would point.

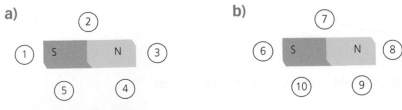

2 Sketch the magnetic field around a bar magnet and label which parts of the field are the strongest.

3 Sketch the magnetic field around the Earth and compare it with the field around a bar magnet.

4 Explain why steel lasts longer as a permanent magnet than iron.

Extend

1 Why does a compass placed near a magnet point to the north pole of the magnet instead of the north pole of the Earth?

2 Two bar magnets are identical in size and shape, but one is stronger than the other. How would the magnetic fields of these magnets be different? Draw diagrams to help explain your answer.

3 Predict and draw what you think the magnetic field would look like for the following:

a)

b)

c)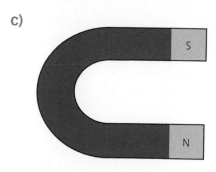

4 Research two different methods of making a permanent magnet from a magnetic material.

5 Explain why repulsion is the true test for magnetism.

6 'Domain theory' is the name given to one way in which scientists try to explain why some materials are magnets and others are not, which imagines that each atom acts like a tiny magnet.

a) Draw a diagram to show how the 'magnets' in each atom would be aligned in a magnet compared with a non-magnetic material.

b) How do you think a magnetic material can be made into a magnet, using the ideas of this theory and some research.

c) Hitting or heating up a magnetised magnetic material can cause it to lose its magnetism. Why do you think this is?

5 Work

» Pushing and pulling

Worked example

Charlie walks up two different flights of stairs. Both are 5 m tall, but flight A has 20 shallow steps whilst flight B has 10 steeper steps. Which flight of stairs requires more work to be done to climb it?

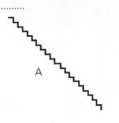

This is a bit of a trick question. As both flights have a vertical distance of 5 m, they both require Charlie to do the same amount of work to climb them – it does not matter how many steps there are.

Know

1 What does work mean in physics?

2 How can you increase the amount of work done on an object?

Apply

1 State three situations where:

   a) you do work

   b) work is done on you.

2 Lifting an apple 1 m vertically requires approximately 1 J of work.

   a) How much work would be done if two apples were lifted instead?

   b) How much work would be done if the single apple was lifted 3 m?

   c) How much work would be done if four apples were lifted 2 m?

   d) How could you do 6 J of work with these apples?

3 Imagine you are lifting up an object. What would happen to the amount of work you have done if:

   a) you lifted the object twice as high

   b) you halved the weight of the object

   c) you lifted the object up and then held it there

   d) you went to the Moon and lifted the object up there

   e) you lifted the object half as high, but it weighed four times as much?

Extend

1 Compare the use of the word 'work' in physics with its use in everyday life. Can you think of any other words that have slightly different meanings in real life like this?

2 Lifting a rucksack onto your shoulders involves doing work on the rucksack against gravity, but walking with it on your back does not. Explain why.

3 Neil Armstrong did less work when lifting the American flag up on the Moon than he did when he practised it on Earth. Explain why, and predict where he could travel to do more work than he would on Earth for the same task.

» Simple machines

Worked example

Explain why it is easier to undo a tightened bolt using a long spanner rather than a short one.

A spanner is an example of a lever, which is a type of simple machine. When you undo a bolt, you are doing work, which follows the equation:

work = force applied × distance to the pivot

Here, the pivot would be around the bolt. Therefore, for a long spanner, the distance to the pivot would be larger, so less force would be needed to do the same amount of work as the short spanner.

Know

1 What is a simple machine?

2 State three examples of simple machines.

3 How do wheels make work easier?

4 What is the difference between an input and output force?

5 What is a lever?

6 What is a fulcrum?

Apply

1 Compare distance and displacement.

2 Describe three situations in which you use a lever to help you do work.

Extend

1 Pulleys were often used in mills to lift the heavy bags of flour up to the top floor. If a sack of flour weighed 100N, and the height to be lifted was 10m, how much force would a miller need to pull with and how far would they need to pull if they used:

   a) one pulley   b) two pulleys   c) four pulleys?

2 Look at the drawing of someone turning a wrench.

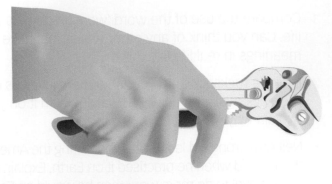

a) Where would the input force be?

b) Where would the output force be?

c) Compare the size of the input and output forces.

d) How could the output force be made smaller?

e) How could the output force be made twice as large?

3 Door handles are usually placed on the opposite side from the hinge of the door, which acts as a fulcrum. Using ideas of levers and work, explain why door handles are placed here, rather than on the side closest to the hinge.

» Calculating work

Worked example

a) How much work is required to lift a box weighting 20 N onto a table that is 0.5 m above the ground?

b) How much extra work needs to be done to push the box 1 m across the table, if there are no frictional forces acting?

a) For this, we need to use the equation:

work = force × distance moved in the direction of the force

The force is the weight of the box, 20 N, whilst the distance is 0.5 m. Therefore:

work = 20 × 0.5 = 10 J

b) This is a bit of a trick question! If there are no frictional forces, then the object is not doing any work as it moves horizontally across the table. The object's weight is a vertical force and so – as the box is not changing height at all – does not need to be taken into account.

Know

1 What is the equation that links work, force and distance?

2 Copy and complete the table below.

Work done (J)	Force (N)	Distance moved in the direction of the force (m)
	5	10
	3	7
	100	0.2
	2.5	500
50	4	
64	8	
1000	20	
660		33
0.8		1.6
300		6

Apply

1 A car's engine works against 150 N of friction as it travels 200 m along a road. How much work is done?

2 How much work is done lifting a 3 N book onto a table 1.5 m above the floor?

3 How much work is done as a 0.1 N coin falls 300 m from the top of a skyscraper?

4 A girl does 20 J of work when she lifts a box onto a shelf 75 cm above the ground.

   a) What is the weight of the box?

   b) What is the mass of the box?

   c) How much work would be done if the box fell to the floor?

   d) What would the difference be between the energy transferred when lifting the box compared with when it was falling?

Extend

1 A boy drops a 100 g ball to the floor. 8 J of work are done.

   a) From what height is the ball dropped?

   b) How much work would be done if it were dropped from twice the height?

   c) What mass would the ball need to be to do 10 J of work if it fell 5 m?

2 A 500 g box is pushed up a slope, as shown in the diagram below.

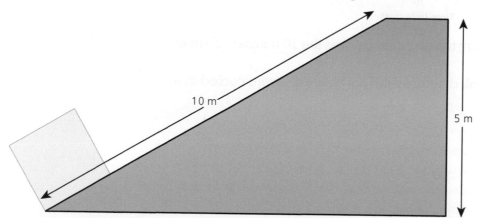

10 m
5 m

   a) Ignoring friction, how much work must be done to push the box to the top of the slope?

   b) How much work must then be done to keep the box still at the top of the slope?

   c) If friction were not ignored, would more or less work need to be done to get the box to the same point on the slope? Explain why.

   d) If the slope was still 5 m tall but shorter in length and so steeper, would it take more, less or the same amount of work to push the box to the top of it? Why?

6 Heating and cooling

» Temperature changes

Worked example

A cup of hot chocolate is left out on a table to cool. Its temperature is taken every 2 minutes and the following graph is produced.

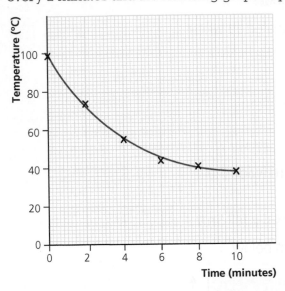

a) How did the rate of energy transfer change over the 10 minutes? Explain your reasoning.

b) How would the graph look different if the hot chocolate were cooled in a fridge?

a) The rate of energy transfer slows down, because the temperature decreases more slowly over time; there is a bigger difference in temperature between readings at the start compared with at the end.

b) If the hot chocolate were cooled in a fridge, the energy would transfer away more quickly and the temperature would decrease more quickly. Therefore, the curve on the graph would be steeper.

Know

1 What is the temperature of an object, a measure of?

2 What is the thermal store of energy, a measure of?

3 How do the particles in a solid change their motion when the solid is heated up?

4 Describe how the motion of particles in a liquid would change as it cooled down.

Apply

1 Look at the table below.

Object	Temperature (°C)
A	10
B	34
C	–5
D	26
E	0
F	–14

a) Which object has the biggest store of thermal energy?

b) In which object will the particles be moving the least quickly? Why?

c) Which two objects would have the most energy flow between them if they were put in contact? Why?

2 Putting an ice cube in a drink causes the liquid to cool down.

a) Why is this, in terms of energy transfers?

b) When will the drink stop cooling down?

3 A hot cup of tea is put in a cold room.

a) What will happen to the temperature of the tea?

b) What will happen to the temperature of the room?

c) Which way does energy flow?

d) What would be different if the tea was put in an insulating cup?

Extend

1 'Close that door, you're letting the cold in.'

a) What is wrong with this sentence?

b) How could you correct it?

2 Ice cream will melt more quickly if left out in the kitchen rather than in a fridge. Why is this? Explain in terms of energy transfer and temperature change.

» Conduction

Worked example

Most saucepans are made of metal, with handles made of either plastic or wood. Why?

Metal is a conductor and allows heat to travel through it easily. This is important for the pan so that the heat from the hob can travel to the contents of the pan. Plastic and wood are insulators, and so do not allow heat to travel through them easily. This is important for the handle so that whoever is using the pan does not get burned.

Know

1 What is conduction?

2 Give three examples of:

   a) good thermal insulators

   b) good thermal conductors.

Apply

1 Good thermal conductors are always bad thermal insulators.

   a) Is this statement always true? Why?

   b) Is the opposite true? Why?

2 Which is most likely to be the best conductor – a solid, liquid or a gas? Why?

3 A hot drink is originally 80°C. It is poured into three different cups. After 10 minutes the drink in cup A is 60°C, the drink in cup B is 65°C and the drink in cup C is 50°C.

   a) Which cup has the biggest energy transfer?

   b) Which cup is made of the best insulating material?

   c) If one cup is made of tinfoil, one is made of Styrofoam and one is paper, which do you think is which? Why?

Extend

1 A conduction star is a piece of equipment that can compare the conductive properties of metals. Look at the diagram of an example below. Each point of the star is made of a different metal: brass, copper, nickel, aluminium and iron. A drawing pin is attached to the end of each point with some petroleum jelly, and the centre of the star is held over a Bunsen burner flame.

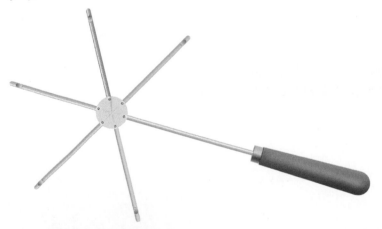

   a) Predict which drawing pin will drop first.

   b) Explain why you think this is.

   c) Research the correct answer.

   d) Discuss how the conduction process would be different between the best and worst conducting metal.

   e) Discuss how the conduction process would be different if one of the points was made of plastic.

» Convection

Worked example

Kettles boil water by setting up convection currents. Make a flow chart to explain how they work, using a diagram to help.

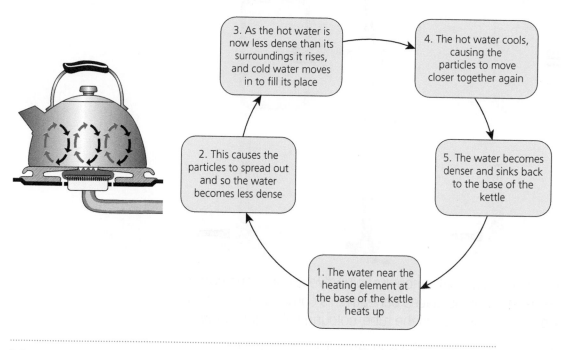

3. As the hot water is now less dense than its surroundings it rises, and cold water moves in to fill its place

4. The hot water cools, causing the particles to move closer together again

2. This causes the particles to spread out and so the water becomes less dense

5. The water becomes denser and sinks back to the base of the kettle

1. The water near the heating element at the base of the kettle heats up

Know

1 What is convection?

2 What happens to the density of a fluid when it is heated?

3 What is the equation for density?

4 Why do hot air balloons float in the air?

5 In a convection current, what happens to:

   a) a warm fluid

   b) a cold fluid?

6 Put these sentences in order, to describe how convection takes place when heating water in a pan.

   A The water rises.

   B The water at the base of the pan gets hotter.

   C As it moves away from the base, it cools down.

   D It becomes denser and sinks again.

   E This causes the water particles to spread out and the water becomes less dense.

Apply

1 A convection tube is a piece of equipment that demonstrates convection currents in the lab.

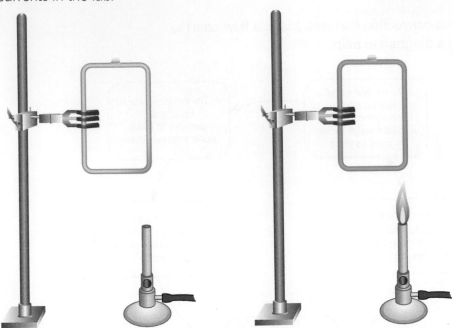

As shown in the diagram, a small amount of blue dye is put in one corner of the tube. A Bunsen burner is placed beneath this corner and turned on. The dye rises up and travels around the tube, colouring all of the water.

a) Why does the dye rise up when the Bunsen burner is turned on?

b) Why does it fall back down again when it reaches the opposite corner?

Extend

1 'Heat rises.' Evaluate the accuracy of this statement.

2 Lava lamps work via convection. A light bulb is placed at the base and switched on, causing the liquid to move.

a) Explain how the lava lamp works.

b) Predict whether or not the lamp would still work if an energy-saving light bulb was used. Explain your reasoning.

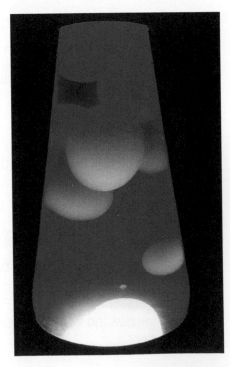

» Radiation

Worked example

Two children are playing on a sunny day. Annie is wearing a white t-shirt, whilst Peter is wearing a black one. Who will feel the warmest, and why?

Peter will feel warmer than Annie, because black is a better absorber of heat than white, and white is a better reflector of heat than black. Therefore, Peter's black t-shirt will absorb more of the sunlight's energy, whilst Annie's t-shirt will reflect more of it.

Know

1 What is radiation?

2 How do we know that radiation can travel through a vacuum?

3 Give three examples of when heat is transferred by radiation.

Apply

1 Why is it more comfortable to wear light colours on a hot day than dark ones? Explain using ideas about radiation.

2 Radiators are usually painted white, but it would be more sensible not to.

   a) What colour should they be painted to emit the most radiation possible?

   b) Why do you think this is not done?

3 A Leslie cube is a piece of equipment that shows how different colours and textures emit heat by radiation differently. It is a metal box with sides made out of different material. It is filled with hot water, and the temperature a certain distance away from each side is measured.

   A set of results from such an experiment is shown below:

Surface	Temperature 5 cm away (°C)
Black, shiny	50
Black, matte	48
White, shiny	45
Copper	47

   a) Which surface is the best emitter of heat?

   b) Which is the worst emitter of heat?

   c) What would you predict the temperature of a white, matte surface to be?

Extend

1 Compare and contrast conduction, convection and radiation as methods of heat transfer.

2 Vacuum flasks use conduction, convection and radiation to keep hot drinks hot and cold drinks cold. Research how.

7 Wave effects

» Sounds and explosions

Worked example

While filming an action movie, a film crew needs to set up an explosion.

a) What type of waves will this explosion cause to be emitted?

b) How would these waves be different if the explosion was bigger?

c) What safety procedures might the cast and crew follow to reduce their risk of harm?

a) An explosion would cause pressure waves, including sound waves. It would also cause light and infrared (heat) waves to be emitted.

b) The sound and pressure waves would have a bigger amplitude, as they would carry more energy. The explosion might also produce more light and heat waves.

c) They need to be as far away as possible. This is because the energy from the explosion spreads out – the further away they are, the 'weaker' the waves will be.

Know

1 State three examples of waves.

2 What is a wave?

3 What do waves transfer?

4 State five keywords that can be used to describe waves.

5 State two things that absorption of a wave can cause.

6 Apart from absorption, what else can waves do when they encounter an obstacle?

7 What is a medium?

8 What is a pressure wave?

9 State some causes of pressure waves.

10 What does a microphone convert sound waves into?

11 What does a loudspeaker convert electrical signals into?

Apply

1 Describe how sound travels from a source to a detector through a medium.

2 Compare microphones and loudspeakers.

3 How would the current produced by a loud sound in a microphone be different from that produced by a quiet sound?

4 Why is it hard to measure the wavelength of sound? Give two reasons.

5 How can the speed of sound be measured?

6 Research other types of pressure wave, and what they can be used for.

Extend

1 Explain how sound can be used to clean something. Why might this method be better than others for delicate objects?

2 Thunder and lightning occur together, but the thunder is usually heard later than the lightning can be seen.

   a) Why is this?

   b) What can be calculated from the time difference between the two?

   c) A storm that is 1.7 km away causes a time difference of 5 seconds between the lightning being seen and the thunder being heard. Use this to calculate the speed of sound.

   d) How far away would the storm be if the time difference was 8 seconds?

   e) What would be the time delay for a storm that was 500 m away?

» Nearly visible

Worked example

Visible light is part of a family of waves called the electromagnetic spectrum. These waves have different wavelengths and frequencies, which cause them to have different properties. A diagram of this family is shown below.

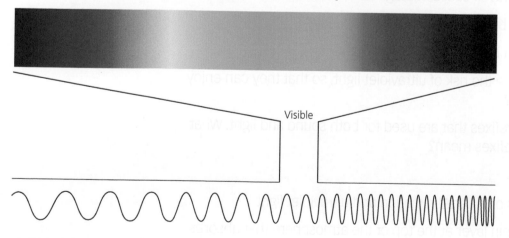

Visible

a) Label on the diagram where ultraviolet light and infrared radiation would be situated.
b) Which of these has the longest wavelength?
c) Which of these has the highest frequency?
d) Which of these can cause damage when living cells absorb it?
e) How can the other types cause damage to humans?

a) Ultraviolet light is light that is 'too violet' for humans to see, so it must go just past the violet end of the visible spectrum. Likewise, infrared radiation is light that is 'too red' for humans to see, and so must be situated just past the red end of the visible spectrum, like so:

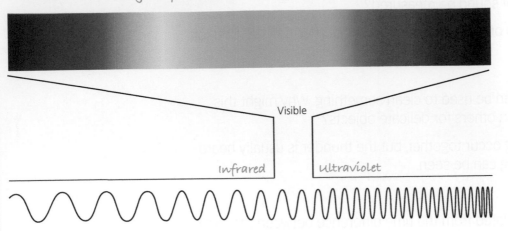

b) Infrared radiation

c) Ultraviolet light

d) Ultraviolet light

e) Infrared radiation causes burns – it is another word for heat. Visible light can damage eyes, if you look at a bright source for too long.

Know

1 What are ultrasound and infrasound?

2 What are ultraviolet light and infrared radiation?

3 What types of radiation does the Sun emit?

Apply

1 Compare sound and light waves.

2 When can ultraviolet waves cause damage to humans?

3 How can this damage be increased?

4 Compare the risks and benefits of ultraviolet light.

5 How can people reduce the risk of ultraviolet light, so that they can enjoy being in the sunshine?

6 Infra- and ultra- are prefixes that are used for both sound and light. What do you think these prefixes mean?

Extend

1 Compare the dangers of infrared radiation and ultraviolet light.

2 The ozone layer is a thin layer at the top of the atmosphere that absorbs most of the ultraviolet light produced by the Sun.

   a) Why is this useful for humans?

   b) The ozone layer has been damaged by the use of certain chemicals called CFCs; it is thinner and now has some holes in it. Predict what effects this might have.

   c) Research what is being done to stop this happening further.

3 Many people sunbathe, despite knowing the risks of ultraviolet light. Discuss why this might be.

4 Visible light, infrared radiation and ultraviolet light are all members of the electromagnetic spectrum. Do some research to answer these questions:

a) What are the other four types of radiation in the spectrum?

b) Which ones are the most dangerous? Why?

c) Which ones have the lowest frequencies?

d) What do all waves in the electromagnetic spectrum have in common?

e) Find a use for each type of radiation.

» Water waves

Worked example

Describe a similarity and difference of water waves and sound waves in air.

Similarity – both water and sound waves are carried by particles. For water, this is water molecules. For sound in air, this is air molecules. Difference – water waves are transverse waves whilst sound waves are longitudinal. This means that the water molecules in a water wave are displaced at right angles to the wave motion – they vibrate up and down – whilst the wave moves sideways. In a sound wave the air particles vibrate parallel to the wave motion – they both move side to side.

Know

1 What is the wavelength of a wave? Draw a diagram to help you explain.

2 What is the symbol for wavelength?

3 What is the amplitude of a wave? Draw a diagram to help you explain.

4 State three methods for using water to make electricity.

Apply

1 What is the difference between transverse and longitudinal waves? Give an example of each.

2 State some advantages and disadvantages of:

a) tidal power   **b)** wave power.

Extend

1 Hydroelectric power plants use dams to produce huge amounts of electricity. Research how hydroelectricity works to answer the following questions.

a) Draw a diagram of a hydroelectric plant.

b) Describe the energy transfers in the system.

c) How could more energy be produced?

d) Give some advantages and disadvantages of hydroelectricity.

8 Wave properties

» Numbers

Worked example

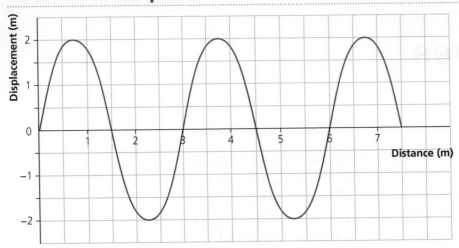

a) Determine the wavelength and amplitude of the wave above.
b) Add a second wave to the diagram, with half the amplitude of the first one.
c) Which of these waves would carry more energy? Why?

a) The wavelength of a wave is the distance between two adjacent identical places on the wave, for example between a peak and the next peak or a trough and the next trough. Here, that would be 3 m. The amplitude is the distance between the midpoint of the wave (here, the x-, or distance axis) and the peak or trough of the wave. Here, that would be 2 m.

b) The amplitude of the new wave should be 1 m, but we need to keep the wavelength the same. Therefore the new wave would look like this:

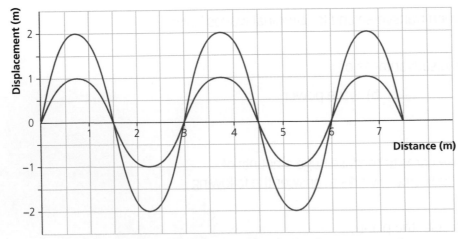

c) The original wave. The energy a wave carries is related to its amplitude and, because the first wave has double the amplitude of the one that we have added, it has more energy.

Know

1 Put these waves in order of speed, from fastest to slowest.

   A Sound in air

   B Light in a vacuum

   C Ripples in a pond

2 Draw a diagram of a wave and label the amplitude and wavelength.

3 Copy and complete the paragraph below using the following key words:

| adjacent | distance | horizontal | energy | ruler | vertical |
| identical | maximum | peak | metres | trough | |

The wavelength of a wave is the _____ between repeating waves.
It is measured on a diagram using a ruler and is the _____ distance
between _____ points on two _____ waves (waves that are next to
each other). The amplitude of a wave is the _____ amount of vibration
of the wave, and is related to how much _____ the wave is carrying. It is
also measured on a diagram using a _____ and is the _____ distance
between the mid-point of the wave and either the top of the _____ or
the bottom of the _____. Both the wavelength and amplitude of a wave
are measured in _____.

Apply

1 Measure the amplitude and wavelength of each of the waves below.

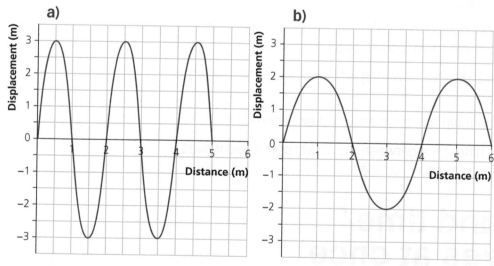

a)

b)

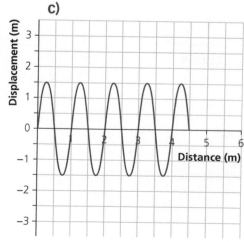

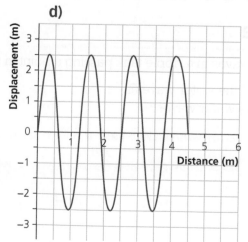

c)

d)

2 What's wrong with the following diagrams?

a)

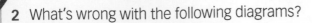

Wavelength

b)

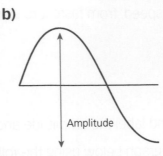

Amplitude

c)

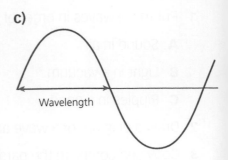

Wavelength

3 A water wave travels 50 km in 20 seconds. What is its wave speed?

4 Sound travels at 340 m/s in air.

  a) How long would it take a noise to travel 1 km?

  b) How far would a noise travel in 2 minutes?

Extend

1 Sound travels at 340 m/s in air.

  a) Mahmoud is standing near a cliff. He claps and hears the echo 8 seconds later. How far away is the cliff?

  b) In a storm, Lucy hears thunder 15 seconds after she sees lightning. How far away must the storm be?

2 Earthquakes are mainly caused by two types of seismic wave – P waves (faster) and S waves (slower).

  a) The epicentre of an earthquake is 500 km away from a town. Which wave type would reach the town first – P or S?

  b) The P waves travel at 25 km/s – how long would they take to reach the town?

  c) The S waves reach the town 20 s later. What is their speed?

  d) What would the time difference between the arrival of the two waves be if the town was 1000 km away instead?

» Waves and time/ two waves at once

Worked example

A student counts 20 water waves in a ripple tank passing a point in 5 seconds.

a) What is the frequency of the water waves?
b) What is the period of the water waves?

a) Frequency is the number of waves that pass a point per second.

If 20 water waves pass in 5 seconds, then $\frac{20}{5} = 4$ waves must pass in 1 second.

Therefore, the frequency of the water waves is 4 Hz.

b) Period is how long it takes for one complete wave to pass a point.

If four waves pass a point per second, then it must take $\frac{1}{4} = 0.25$ seconds for each wave to pass the point.

Therefore, the period of the wave is 0.25 seconds.

Know

a) What is meant by the 'frequency' of a wave and what is it measured in?

b) What is meant by the 'period' of a wave and what is it measured in?

c) How are frequency and period related to each other?

Apply

1 A radio wave has a frequency of 3 000 000 Hz. What does this mean in simple terms?

2 The musical note 'middle C' has a frequency of about 260 Hz.

    a) How long would it take for 1000 'middle C' sound waves to pass a point?

    b) What would the period of 'middle C' sound waves be?

3 Put the following waves in order of period:

    Wave A   frequency of 100 Hz

    Wave B   frequency of 50 Hz

    Wave C   frequency of 200 Hz

4 If the following waves were travelling in opposite directions and met, they would interfere with each other. Match up the diagrams to show how.

1 +

A _____

2 +

B

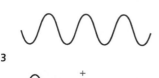

3 +

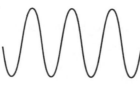

C

Extend

1 What would the period be for a wave of frequency 1 kHz?

2 Wave 1 has a frequency that is twice that of wave 2. What could you say about the relationship between their periods?

3 This pattern is produced when light waves from two sources meet and interfere:

   a) Use ideas of superposition to discuss why the bright and dark bands might occur.

   b) Why do you think there is a fuzzy edge? What could be happening to the light beams as they meet there?

» The wave equation and light waves

Worked example

A guitarist strums a string on her guitar. It produces a sound of frequency 440 Hz.

a) Assuming that the speed of sound in air is 340 m/s, what must the wavelength of the sound be?

b) How would the wavelength change if the frequency was doubled?

a) To answer this question, we need to use the wave equation:

   wave speed = frequency × wavelength

This needs to be rearranged in terms of wavelength:

$$\text{wavelength} = \frac{\text{wave speed}}{\text{frequency}}$$

We can now put the numbers in:

$$\text{wavelength} = \frac{\text{wave speed}}{\text{frequency}} = \frac{340}{440} = 0.77\,\text{m}$$

b) If the frequency was doubled, the wavelength would halve, because the speed of sound would stay the same:

$$\text{wavelength} = \frac{\text{wave speed}}{\text{frequency}} = \frac{340}{880} = 0.39\,\text{m}$$

Know

1 What is the wave equation?

2 What is the speed of light?

3 Why is standard form often used when writing down the speed of light?

4 What can one thousand-millionth of a metre also be known as?

Apply

1 What would the speed of a wave be if it had:

   a) a frequency of 10 Hz and wavelength of 0.2 m

   b) a wavelength of 10 m and a frequency of 75 Hz

   c) a frequency of 0.3 Hz and wavelength of 2 km

   d) a wavelength of 60 m and a frequency of 10 kHz?

2 What would the frequency of a wave of wavelength 2 m be if it was travelling at 3 km/s?

3 What would the wavelength of a 2 kHz wave travelling 5 000 000 m/s be?

4 Why is the speed of light given the symbol c?

Extend

1 A seismic wave travels 5000 km in 2 hours.

   a) What is its speed?

   b) If it has a wavelength of 4 m, what must its frequency be?

   c) What would the period of this seismic wave be?

   d) As the wave travels deeper into the Earth, it gets faster. If its frequency stays the same, what must happen to its wavelength?

2 All electromagnetic waves travel at the same speed, 3×10^8 m/s, but have different wavelengths and frequencies. Use the wave equation to calculate the missing typical values for each type in the table below.

Electromagnetic radiation	Frequency (Hz)	Wavelength (m)
Gamma rays		3×10^{-13}
X-rays	1×10^{19}	
Ultraviolet light		3×10^{-9}
Visible light	1×10^{15}	
Infrared radiation		3×10^{-5}
Microwaves	1×10^{10}	
Radio waves		3

9 Periodic table

» Structure of the periodic table

Worked example

Explain why a balloon filled with helium floats in air, but a balloon filled with xenon sinks.

The density of the helium is lower than the density of air, whereas the density of xenon is greater than the density of air.

Know

1 State what is meant by each of the following terms:

a) element

b) periodic table

c) a group in the periodic table

d) a period in the periodic table

2 The noble gases are described as inert elements. State what is meant by 'inert'.

3 Give one use for each of the following noble gases. Choose a different use for each gas.

a) helium

b) neon

c) argon

Apply

1 The table below gives some properties of the noble gases.

Name of noble gas	Boiling point (°C)	Density (g/dm³)
Helium	−269	0.18
Neon	−246	0.90
Argon	−186	1.78
Krypton	−152	
Xenon		5.9
Radon	−62	9.7

a) Use the information in the table to state the relationship between:

i) the boiling point and the position of the noble gas in the periodic table

ii) the density and the position of the noble gas in the periodic table.

b) Use the information in the table to estimate a value for:

i) the boiling point of xenon

ii) the density of krypton.

c) Explain which balloon – one filled with xenon or one filled with radon – would fall to the ground the more quickly when released.

Extend

1 Place the following elements in their correct group in the periodic table. (Note that the letters used are *not* the symbols for the elements.)

   A A solid that can be cut with a knife to expose a shiny surface that quickly turns dull. The solid reacts violently with water to form a strongly alkaline solution.

   B A gas that forms no chemical compounds.

   C A gas that reacts with sodium to form a salt.

2 Use a data book or the internet to find the density of each of the following metals: barium, calcium, copper, iron, magnesium, sodium, strontium and potassium.

   a) Use the values of the densities to place these metals into three groups, labelled A, B and C. State the reasons for your choices.

   b) Your groups should correspond to areas of the periodic table. State the area of the periodic table in which you would find each of the groups you have labelled A, B and C.

   c) State the trend in the densities of the group 2 metals magnesium, calcium, strontium and barium.

» Group 1 (the alkali metals)

Worked example

State which alkali metal reacts most vigorously when heated in oxygen? Justify your answer.

Caesium reacts most vigorously. The reactivity of the alkali metals increases as you descend the group and caesium is at the bottom of the group.

Hint

Remember how the reactivity of the alkali metals changes as you go down the group. Does the reactivity increase or decrease as you go down the group?

Know

1 How does the reactivity of the alkali metals change as you go down the group?

2 Why do the alkali metals need to be stored in oil?

3 Name the gas produced when sodium reacts with water.

4 Name the compound formed when lithium reacts with oxygen.

Apply

1 When a small piece of lithium is added to water it fizzes and eventually disappears forming a solution.

   a) Write a word equation for the reaction that takes place between lithium and water.

   b) State and explain the effect that the solution formed has on litmus paper.

   c) State two similarities and two differences between the reactions of lithium and potassium with water.

2 This question is about the five group 1 metals; lithium (Li), sodium (Na), potassium (K), rubidium (Rb) and caesium (Cs).

Identity the substances **A**, **B**, **C**, **D**, **E** and **F** from the following descriptions.

a) Metal **A** has the lowest density of the five group 1 metals.

b) When the group 1 metal **B** is added to water it forms a small ball and moves around the surface of the water very quickly. A lilac flame is also seen. A gas **C** is given off. A white trail dissolves in the water to form a solution of compound **D**.

c) When the group 1 metal **E** is heated in oxygen it burns with a yellow-orange flame to form a white solid compound **F**.

Extend

1 A small piece of lithium is added to a large volume of water in a trough. The lithium reacts with the water and the following observations are made:

- The lithium moves around on the surface of the water.

- Bubbles of gas are seen.

- The piece of lithium gets smaller and smaller and eventually disappears.

- The solution formed turns red litmus paper blue.

a) Predict how the reaction of rubidium with water compares with the reaction of lithium with water.

b) Write a word and a chemical equation for the reaction that takes place between rubidium and water.

c) Suggest a value for the pH of the solution formed when rubidium reacts with water.

» Group 7 (the halogens)

Worked example

There are five halogens in group 7 of the periodic table. These are fluorine, chlorine, bromine, iodine and astatine. All react with sodium when heated.

State which halogen reacts least vigorously when heated with sodium. Justify your answer.

Astatine reacts least vigorously with sodium. The reactivity of the halogens decreases as you go down the group. Astatine is the least reactive since it is at the bottom of the group 7.

> **Hint**
>
> Remember how the reactivity of the halogens changes as you go down the group. Does the reactivity increase or decrease as you go down the group?

Know

1 What is the meaning of the word 'halogen'?

2 How does the reactivity of the halogens change as you go down the group?

3 What is the name of the compound formed when sodium reacts with chlorine?

Apply

1 The table gives information about the first four elements in group 7 of the periodic table.

Element	Physical state at 20°C	Colour at 20°C	Name of compound formed when reacted with hydrogen
Fluorine	Gas	Pale yellow	Hydrogen fluoride
Chlorine	Gas	Pale green	Hydrogen chloride
Bromine	Liquid	Red-brown	Hydrogen bromide
Iodine	Solid	Dark grey	Hydrogen iodide

Astatine is the fifth element in Group 7. It is possible to make predictions about astatine by comparison with the other elements in Group 7.

a) Predict the physical state of astatine at 20°C.

b) Predict the colour of astatine at 20°C.

c) Predict the name of the compound formed when astatine reacts with hydrogen.

2 The table shows some information about the halogens in group 7 of the periodic table.

Halogen	Density (g/cm³)	Melting point (°C)	Boiling point (°C)	Atomic radius in pm
Fluorine	0.0017	−220	−188	72
Chlorine	0.0032	−101	−34	99
Bromine	3.1	−7		114
Iodine		114	184	133
Astatine	6.4	302	337	

a) Estimate a value for the boiling point of bromine.

b) Estimate a value for the density of iodine.

c) Estimate a value for the atomic radius of astatine.

d) State the relationship between the reactivity of the halogens and the atomic radius.

e) Explain why the densities of fluorine and chlorine are much lower than those of bromine, iodine and astatine.

Extend

1 A halogen will displace a less reactive halogen from a solution of its metal halide. The chemical equation for the displacement of halogen Y by halogen X is:

$$X_2(aq) + 2NaY(aq) \rightarrow 2NaX(aq) + Y_2(aq)$$

The colours of solutions of the halogens are:

- chlorine solution – pale green

- bromine solution – orange

- iodine solution – brown.

You are supplied with the following six solutions:

- sodium chloride solution
- sodium bromide solution
- sodium iodide solution
- chlorine solution
- bromine solution
- iodine solution

a) Describe two experiments you could perform to show that the order of reactivity of the halogens is chlorine > bromine > iodine. In your description include the observations you would make.

b) Write chemical equations for the reactions that take place in your experiments.

10 Elements

» Representing elements and compounds

Worked example

What is the difference between a mixture of elements and a compound?

In a mixture the elements are not chemically bonded together and can be separated using appropriate physical processes (such as filtration, evaporation, distillation or chromatography).
In a compound the elements are chemically bonded together and can only be separated by chemical reactions.

Know

1 State what is meant by each of the following terms:

    a) element

    b) atom

    c) molecule

    d) compound

    e) polymer

Apply

1 State two ways in which a mixture of hydrogen and oxygen differs from water.

> **Hint**
>
> Water has the chemical formula H_2O.

2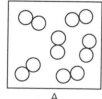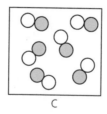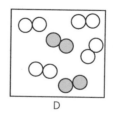

    A   B   C   D

    Which of the diagrams A, B, C and D represent:

    a) a pure element

    b) a pure compound

    c) a mixture of two elements

    d) a mixture of an element and a compound?

3 What can you deduce about the two substances E and F shown in the following diagrams?

    E

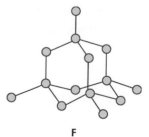

    F

Extend

1 The diagrams show two different forms of carbon. Both forms are giant structures.

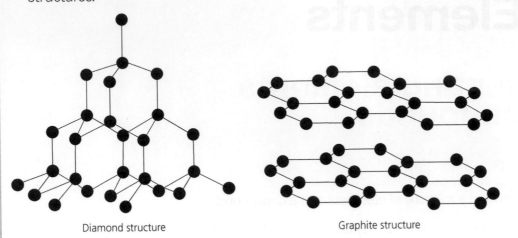

Diamond structure

Graphite structure

Diamond is very hard and can be used as an abrasive. Graphite is soft and can be used as a lubricant.

a) Suggest why diamond is hard.

b) Suggest why graphite is soft and can be used as a lubricant.

2 The table gives the names and uses of some polymers.

Polymer	Typical use	Properties
Poly(ethene)	Plastic bags	
PVC	Water pipes	
PVC	Outer layer of electric wires	
Nylon	Clothing	
PTFE	Coating on frying pans and saucepans	
Lycra	Sports clothing	

Copy and complete the table by stating the properties of each polymer that make the polymer suitable for the use given.

» Interpreting chemical formulae

Worked example

Give the names of the elements present in the compound with the formula $NaBrO_3$. How many of each type of atom are present?

The compound contains sodium, bromine and oxygen. There is one sodium atom, one bromine atom and three oxygen atoms.

Know

1 Give the chemical symbol for each of the following elements:

a) lithium

b) beryllium

c) boron

d) fluorine

e) neon

2 Give the names of the elements that have the following symbols:

a) Si

b) P

c) Ca

d) H

e) He

3 Give the chemical formula for each of the following elements:

a) hydrogen

b) oxygen

c) chlorine

d) sulfur

e) fluorine

Apply

1 Give the name of each of the following compounds:

a) KCl

b) Li_3N

c) Na_2CO_3

d) $MgSO_4$

e) $Al(NO_3)_3$

2 Give the formula for each of the following compounds:

a) carbon monoxide

b) carbon dioxide

c) sulfur trioxide

Extend

1 Copy and complete the following table by inserting the correct formulae for the missing compounds. Use the patterns in the formulae to work out the formulae of the missing ones.

	Chloride	Oxide	Sulfate	Hydroxide	Nitrate
Sodium	NaCl		Na_2SO_4		$NaNO_3$
Potassium		K_2O		KOH	
Magnesium	$MgCl_2$	MgO	$MgSO_4$		
Calcium				$Ca(OH)_2$	$Ca(NO_3)_2$
Aluminium	$AlCl_3$		$Al_2(SO_4)_3$		

Hint

Be careful when writing symbols. Always print the letters – do not use joined-up writing. If the symbol has two letters, the first letter must be upper case (i.e. a capital letter) and the second must be lower case.

Hint

Make sure that you write any numbers as subscripts. The formula for bromine is Br_2. Br2 and Br^2 are both incorrect.

Hint

Sodium and potassium are both in group 1; magnesium and calcium are both in group 2; aluminium is in group 3.

11 Chemical energy

» Exothermic and endothermic reactions

Worked example

Sherbet powder is a mixture of sodium hydrogencarbonate and tartaric acid.

When a student put some sherbet powder into his mouth he noticed that the mixture fizzed and also that his mouth became cold. Explain these two observations.

When sherbet powder mixes with water a reaction takes place.
The fizzing is caused by carbon dioxide gas being formed.
The reaction mixture gets cold because the reaction is endothermic;
it is taking in heat energy from the student's mouth.

Know

1 State what is meant by each of the following terms:

   a) exothermic reaction

   b) endothermic reaction

2 What happens to the temperature of a reaction mixture during an exothermic reaction?

Apply

1 When sodium hydroxide solution is added to dilute hydrochloric acid a reaction takes place. Describe how you could find out if the reaction is exothermic or endothermic.

2 The following two graphs show how the temperature of a reaction mixture changes during the course of the reaction.

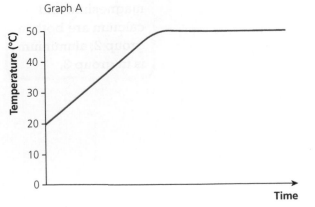

Graph A

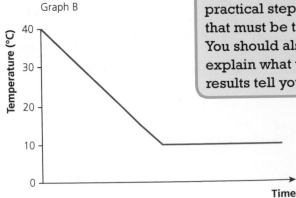

Graph B

> **Hint**
>
> When asked to comment on two different observations, give your answer to each in a separate sentence. The fizzing is caused by a gas – you should give the name of this gas in your answer. If the student's mouth is getting cold then heat energy must be transferred from his mouth to the reaction mixture. What does this tell you about the reaction?

> **Hint**
>
> When answering a question in which you have to describe an experiment, it is important to give all of the essential practical steps that must be taken. You should also explain what your results tell you.

a) Does Graph A show an endothermic or an exothermic reaction? Justify your choice.

b) Does Graph B show an endothermic or an exothermic reaction? Justify your choice.

Extend

1 Dynamite is an explosive compound. It contains the elements carbon, hydrogen, nitrogen and oxygen. When it explodes it produces nitrogen gas, carbon dioxide gas, water (in the form of steam) and oxygen.

a) Write a word equation for the explosion of dynamite.

b) Dynamite can be made to explode by detonating it. This involves setting off a small explosion of another substance using an electric current. Why is it necessary to detonate the dynamite before it will explode?

c) Suggest another way of detonating dynamite.

d) Why does the explosion of dynamite cause so much damage to materials around it?

2 Photosynthesis is a process in which glucose ($C_6H_{12}O_6$) is made from carbon dioxide and water. The chemical equation for this reaction is:

$$6CO_2 + 6H_2O \rightarrow C_6H_{12}O_6 + 6O_2$$

It is an endothermic reaction and the overall energy change for the reaction is +1257 kJ/mol. The energy level diagram for this reaction is shown on the right.

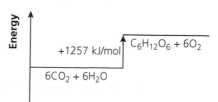

a) Respiration can be considered to be the reverse of photosynthesis. Write a chemical equation to represent respiration.

b) Copy the energy level diagram shown above for photosynthesis and draw alongside it, and to the same scale, the energy level diagram for respiration.

c) Photosynthesis is an endothermic process. Where does the energy for it come from?

d) State two ways in which the burning of fuels is similar to respiration.

» Explaining energy changes

Worked example

When copper carbonate is heated it decomposes to form copper oxide and carbon dioxide. Fill in the gaps in the following sentences to explain why this reaction is endothermic.

More energy is taken in when the _____ in copper carbonate are broken than is given out when the bonds in _____ and _____ are formed.

More energy is taken in when the <u>bonds</u> in copper carbonate are broken than is given out when the bonds in <u>carbon dioxide</u> and <u>copper oxide</u> are formed.

Know

1 State what is meant by each of the following terms:

   a) catalyst

   b) chemical bond

2 Use words from the box to fill in the gaps in the sentences that follow.

endothermic exothermic given out taken in

   During a chemical reaction energy is _____ when chemical bonds
   are broken and energy is _____ when chemical bonds are made. If
   the energy given out is greater than the energy taken in, the reaction
   is _____. If the energy given out is less than the energy taken in, the
   reaction is _____.

Apply

1 The following energy level diagram represents a reaction in which the
   energy of the reactants is the same as the energy of the products.

   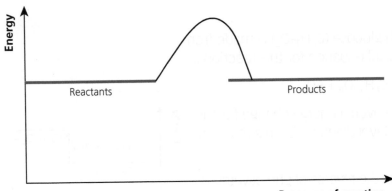

   a) Draw an energy level diagram to represent an endothermic reaction.
   Draw an arrow on the diagram to show the overall energy change of
   the reaction.

   b) Draw an energy level diagram to represent an exothermic reaction.
   Draw an arrow on the diagram to show the overall energy change of
   the reaction.

2 Copy the following energy level diagram for a reaction.

   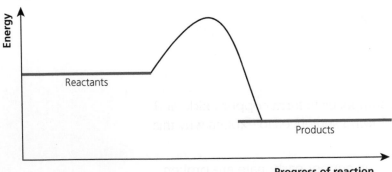

   a) Draw an arrow on your diagram to show the energy taken in to the
   break the bonds of the reactants. Label this arrow A.

   b) Draw a second arrow on your diagram to show the energy given out
   when the bonds of the products are made. Label this arrow B.

3 Copy the following energy level diagram for a reaction.

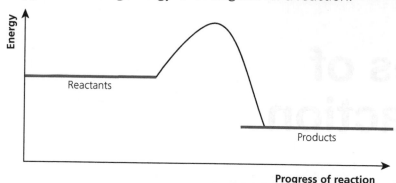

Draw a curve on your diagram to represent the effect of the catalyst on the reaction. Label the curve C.

Extend

1 The table gives some bond energies.

Bond	Bond energy (kJ/mol)
C–H	412
O=O	496
C=O	743
H–O	463

The chemical equation for the reaction between methane (CH_4) and oxygen is:

$$CH_4 + 2O_2 \rightarrow CO_2 + 2H_2O$$

The equation below shows all of the bonds of the reactants and the products of this reaction:

a) Calculate the energy taken in when all of the bonds of the reactants (CH_4 and $2O_2$) are broken.

b) Calculate the energy given out when all of the bonds in the products (CO_2 and $2H_2O$) are made.

c) Use your answers to parts a) and b) to calculate the overall energy change for the reaction.

> **Hint**
>
> The final answer must have either a negative or a positive sign. The reaction between methane and oxygen is exothermic.

12 Types of reaction

» Thermal decomposition

Worked example

A hydrated salt contains water that is chemically bonded to the salt. This water is known as 'water of crystallisation'. When a hydrated salt is heated it loses its water of crystallisation.

The equation for the action of heat on hydrated copper sulfate is:

$$CuSO_4.5H_2O(s) \rightarrow CuSO_4(s) + 5H_2O(g)$$

Explain why this reaction is classified as thermal decomposition.

The hydrated copper sulfate has been broken down into simpler substances by the action of heat.

> **Hint**
>
> You must not use the word 'decomposition' or 'decompose' in your answer. It is necessary to explain what is meant by decomposition.

Know

1 State what is meant by each of the following terms:

   a) reactants

   b) products

   c) thermal decomposition

   d) chemical reaction

   e) physical change

2 Give the state symbol for each of the following physical states:

   a) solid

   b) liquid

   c) gas

   d) aqueous (dissolved in water)

3 What does the law of conservation of mass say about a chemical reaction?

Apply

1 **a)** You are supplied with two test tubes, a delivery tube bent at ninety degrees, a rubber bung and a beaker.

   Draw a labelled diagram, including this apparatus, to show how you would heat a sample of hydrated cobalt chloride, collect the water vapour given off, and condense the vapour to liquid water.

   b) What test would you do to show that the liquid collected is pure water?

   c) How would you find out if the hydrated cobalt chloride gains mass, loses mass or stays the same mass on heating?

   d) What change in mass would you expect? Explain your answer.

> **Hint**
>
> The water vapour will need to be cooled to condense it.

2 Blue crystals of hydrated copper sulfate are dry to the touch. However, when they are heated, water vapour is given off.

The residue left after heating is a white powder. When water is added drop by drop to this white powder, the powder turns blue and gets quite hot.

A student wants to find what proportion of the crystals is water. He heats a weighed sample of hydrated copper sulfate in a crucible until all the water has been removed. He then allows the residue to cool and re-weighs it.

The table shows her results.

Mass of empty crucible	20.4 g
Mass of crucible + hydrated copper sulfate	32.9 g
Mass of crucible + white powder	28.9 g

a) Calculate the mass of hydrated copper sulfate used.

b) Calculate the mass of water given off.

c) Calculate the percentage of water present in the hydrated copper sulfate.

d) How could the student make sure that all of the water had been given off on heating?

e) Why does the white powder get hot when water is added to it?

Extend

1 When some metal carbonates are heated they decompose to form their oxide and give off a gas.

The apparatus shown in the diagram is used to heat a metal carbonate and to pass any gas given off through limewater. If carbon dioxide is given off the limewater will turn milky.

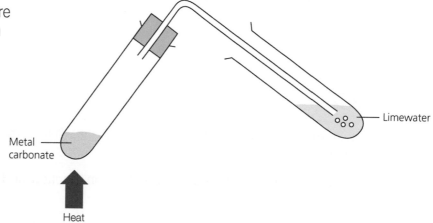

Metal carbonate

Limewater

Heat

The experiment was done with five different metal carbonates.
The time taken for the limewater to go milky was measured in each case.
The results are shown in the table below.

Metal carbonate	Time taken for limewater to go milky (in seconds)
Calcium carbonate	125
Copper carbonate	11
Magnesium carbonate	84
Sodium carbonate	Limewater did not turn milky even after heating for 10 minutes
Zinc carbonate	30

a) Name the gas that turns limewater milky.

b) Suggest why the limewater did not go milky when sodium carbonate was heated.

c) What is the relationship between the reactivity of the metal in the carbonate and time taken for the carbonate to decompose? Explain how you arrived at your answer.

2 When a lump of a mineral called aragonite is heated very strongly it crumbles to form a white powder.

The powder is allowed to cool and then divided into two portions.

A few drops of water are added to one portion. The powder hisses and steam is given off.

The mixture produced is filtered and carbon dioxide is bubbled through the filtrate. The colourless solution turns milky.

Dilute hydrochloric acid is added to the other portion and the mixture is warmed. The solid disappears, but there is no other indication that a reaction is taking place.

Dilute hydrochloric acid is added to another lump of aragonite. Carbon dioxide gas is given off.

a) Do you think that it was just the heating alone that made the lump of aragonite crumble, or is there a better explanation?

b) What kind of reaction was taking place when water was added to the white powder formed on heating aragonite? Explain your reasoning.

c) Give the name of the solution that turned milky when carbon dioxide was bubbled through it.

d) What is the name of the compound that is present in aragonite? Explain your reasoning.

» Combustion

Worked example

When magnesium is heated in air it burns with a bright, white flame, forming a white powder.

Explain why this reaction is classified as a combustion reaction, and write a word equation for the reaction.

The magnesium reacts with oxygen to release energy in the form of heat and light:

magnesium + oxygen → magnesium oxide

Know

1 State what is meant by the term combustion.

2 What is a fuel?

Apply

1 Classify each of the following reactions as combustion or thermal decomposition:

a) Sodium burning in oxygen to form sodium oxide.

b) Potassium hydrogencarbonate changing into potassium carbonate, water and carbon dioxide when heated.

c) Methane burning in air to form carbon dioxide and water.

d) Lead nitrate changing into lead oxide, nitrogen dioxide and oxygen when heated.

2　When magnesium burns in air, it reacts with oxygen to form magnesium oxide.

A student wants to find out the mass of magnesium oxide formed when a sample of magnesium burns in air. She uses the following method.

Step 1: Weigh a crucible and lid.

Step 2: Place some magnesium ribbon in the crucible, replace the lid and reweigh.

Step 3: Heat the crucible, as shown in the diagram, until the magnesium burns.

Step 4: Lift the lid from time to time until there is no sign of further reaction.

Step 5: Allow the crucible and lid to cool, and reweigh.

Step 6: Repeat the heating, cooling and reweighing until two consecutive masses are the same.

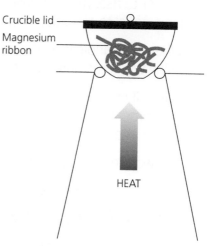

a) i) Suggest why it is necessary to lift the lid from time to time (step 4).

ii) Suggest why it is necessary to repeat the heating until two consecutive weights are the same (step 6).

b) The student records the following results:

Mass of empty crucible and lid	26.7 g
Mass of crucible, lid and magnesium	29.1 g
Mass of crucible, lid and magnesium oxide	30.7 g

i) Calculate the mass of magnesium used in the experiment.
ii) Calculate the mass of magnesium oxide formed in the experiment.

3　The diagram shows methane (natural gas) burning in air, and how the products of the reaction are collected and tested.

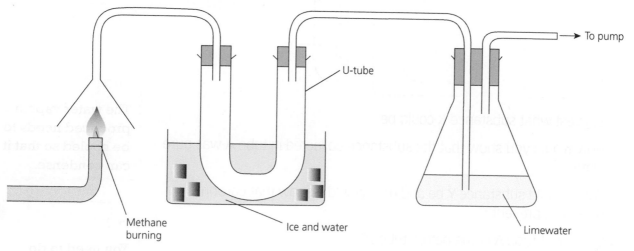

Water collects in the U-tube and the limewater truns milky, showing that carbon dioxide is formed. Suggest what two elements are present in methane.

Extend

1 The apparatus shown is set up in an attempt to show that 21% of the air is oxygen.

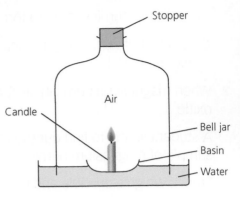

The theory behind the experiment is that the burning candle will use up all of the oxygen in the air and this will cause a decrease in the pressure of gas inside the bell jar. The pressure of the air outside the bell jar will then cause the level of water inside the bell jar to rise. The decrease in the volume of gas inside the bell jar is equal to the volume of oxygen used up.

When the candle is lit, the water level first of all falls. It then rises. It stops rising when the flame goes out.

After allowing the apparatus to cool, the volume of air used up in the experiment is much less than 21%. The experiment has **not** shown that 21% of the air is oxygen.

The word equation for the reaction taking place is:

candle + oxygen gas → carbon dioxide gas + water

a) Why does the water level fall at the beginning of the experiment?

b) Suggest two reasons why the experiment produces an incorrect result.

c) The oxygen used in the burning of the candle produces an equal volume of carbon dioxide. Why then does the water level rise?

2 The apparatus shown in the diagram is used to show that carbon dioxide and water are produced when ethanol is burned.

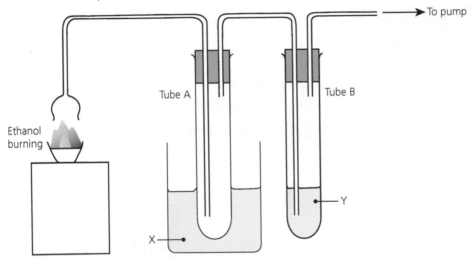

a) Suggest what substance X could be.

b) How would you show that the substance collected in tube A was pure water.

c) What would substance Y be and how would it show that carbon dioxide is present?

d) Why must tube A come before tube B?

e) What would you see in tubes A and B if the experiment were repeated, but with the burning ethanol removed so that only air was drawn through the apparatus?

> **Hint**
> The water vapour produced needs to be cooled so that it can condense.

> **Hint**
> You need to do both a chemical test and a physical test.

13 Climate

» The carbon cycle

Worked example

Describe how burning wood from trees is affecting the carbon cycle.

Wood is a carbon-based fuel. When it burns carbon dioxide is formed and this goes into the atmosphere. Cutting down the trees to produce the wood for burning means that less carbon dioxide is taken out of the air by photosynthesis. The overall effect is that carbon dioxide is added to the atmosphere.

Hint

Consider the effects of both cutting down the trees to supply the wood, and the effect that burning the wood will have.

Know

1 The bar chart shows the percentage by volume of the gases in the Earth's atmosphere.

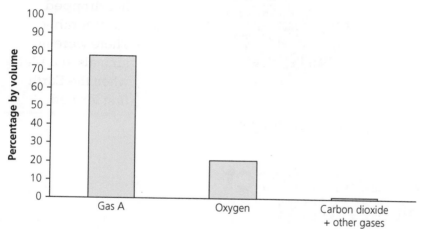

Give the name of gas A.

2 Which of the following gases occurs naturally in the Earth's atmosphere?

A argon

B carbon monoxide

C chlorine

D hydrogen

3 The gases in the Earth's earliest atmosphere were thought to come from which of the following?

A ice caps

B plants

C the ocean

D volcanoes

Apply

1 Apart from burning fossil fuels, describe two other activities that affect the amounts of gases in the atmosphere.

2 Describe how carbon dioxide in the atmosphere becomes calcium carbonate in rocks.

Extend

1 Scientists believe the Earth was formed about 4.5 billion years ago, and that its early atmosphere was probably created from the gases escaping from the Earth's interior. This early atmosphere probably consisted of mostly carbon dioxide and water vapour, with a smaller proportion of nitrogen, ammonia and methane.

The two pie charts show the difference in the composition of the Earth's early atmosphere, around 3.5 billion years ago, and the Earth's atmosphere today.

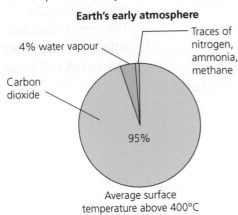

Earth's early atmosphere

Traces of nitrogen, ammonia, methane
4% water vapour
Carbon dioxide
95%
Average surface temperature above 400°C

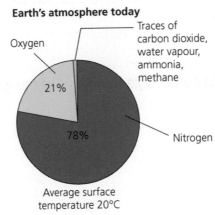

Earth's atmosphere today

Traces of carbon dioxide, water vapour, ammonia, methane
Oxygen
21%
78%
Nitrogen
Average surface temperature 20°C

> **Hint**
>
> Note that the average surface temperature has dropped considerably. There were no animals and plants when the Earth first formed.

Research and then explain why the composition has changed.

» The greenhouse effect and global warming

Worked example

Explain how global warming might affect the ice caps and how this may cause an increase in sea levels.

> **Hint**
>
> The melting of icebergs does not raise the sea level.

As the temperature of the Earth's surface rises, some of the thermal energy will be transferred to the ice caps. The rising temperature may cause more icebergs to form by weakening the glaciers, causing more cracks and making ice more likely to break off. As soon as the ice falls into the ocean, the ocean rises a little.

Know

1 Explain what is meant by each of the following terms:

   a) the greenhouse effect

   b) global warming

   c) weather

   d) climate

Apply

1 Explain how the greenhouse effect warms the Earth's atmosphere.

Extend

1 The graphs show the concentration of carbon dioxide in the atmosphere and the mean global temperatures from 1960 to 2010.

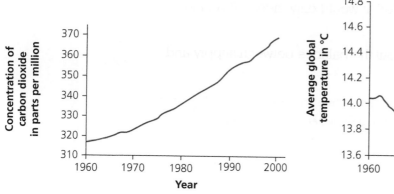 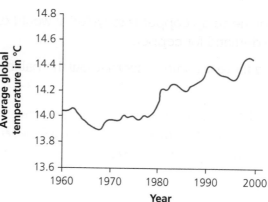

   a) State three ways in which carbon dioxide is released into the atmosphere.

   b) Explain whether these graphs provide evidence that an increase in carbon dioxide levels in the atmosphere is causing the Earth's temperature to rise.

14 Earth resources

» Finite resources

Worked example

The worldwide demand for copper increased steadily from 2 million tonnes in 1900 to 20 million tonnes in 2010. The demand is predicted to continue to increase and could be as high as 40 million tonnes by 2030.

The deposits of copper in the Earth's crust are estimated to be about 289 million tonnes.

Even if all of the scrap copper is recycled, it would only meet 50% of the worldwide demand for copper.

Explain why there is some concern about the balance between supply and demand.

The demand for copper is rising, so copper deposits will eventually run out in the future. Therefore, supply will not meet the demand, even if all of the scrap copper is recycled.

Know

1 Explain what is meant by each of the following terms:

   a) natural resources   **b)** extraction (of a metal)   **c)** recycling

2 **a)** Give the name of the two most common elements found in the Earth's crust.

   b) Give the name of the most common metal found in the Earth's crust.

Apply

1 The table below shows some information about the extraction of copper from three different copper compounds obtained from the Earth's crust.

	Mass of compound (g)	Mass of copper extracted from the compound (g)	Mass of copper extracted per gram of compound (g)	Mass of copper extracted per kilogram of compound (g)
Compound 1	400	20	20/400 = 0.05	50
Compound 2	400	30	30/400 = 0.075	75
Compound 3	400	24		

   a) Calculate the mass of copper extracted per gram for compound 3.

   b) Calculate the mass of copper extracted per kilogram for compound 3.

   c) Which compound produces the least waste material? Explain how you arrived at your answer.

d) Identify which of the following compounds might produce sulfur dioxide when copper is extracted from them:

- $CuCO_3$
- Cu_2O
- CuO
- CuS
- $CuFeS_2$
- Cu_2S

2 A mining company wants to find a new area to mine for metals.

A scientist from the mining company tests rocks from two possible sites. Her results are shown in the table.

Elements present in the rock	Percentage of each element present in the rock from site X	Percentage of each element present in the rock from site Y
Aluminium	14	14
Copper	7	0
Iron	9	11
Oxygen	39	45
Silicon	26	19
Other elements	5	

a) Calculate the percentage of other elements present in the rock from site Y.

b) State the similarities between the rocks found in each site.

c) Give two reasons why site X is the better site for the company to use to mine metals.

Extend

1 The table gives some information about aluminium and tin.

Give two reasons why it could be more important to recycle tin than to recycle aluminium.

Metal	Cost of 1 kg of metal	Amount of metal in Earth's crust
Aluminium	£1.40	8%
Tin	£13.10	0.0002%

2 Research the advantages and disadvantages of recycling metals.

» Extracting metals using displacement reactions

Worked example

The order of reactivity of some metals is listed below. The list also contains the non-metal carbon. The most reactive metal is placed first.

sodium > calcium > magnesium > carbon > zinc > iron > lead > copper

Suggest how zinc could be obtained from zinc oxide.

Zinc oxide is mixed with carbon and the mixture is heated. The following reaction will take place:

zinc oxide + car bon → zinc + carbon dioxide

The metals sodium, calcium or magnesium would also convert zinc oxide to zinc, but these metals are more expensive than carbon.

> **Hint**
>
> Zinc can be obtained by making use of a displacement reaction. Choose a suitable substance that will be will displace zinc from zinc oxide, and will be the least expensive.

Know

1 Metals can be found in the Earth's crust as either minerals or ores. State what is meant, in this context, by the terms:

   a) mineral

   b) ore.

Apply

1 Copper could be obtained by heating copper oxide with either carbon or magnesium. The equations for the two reactions are:

$$CuO(s) + Mg(s) \rightarrow Cu(s) + MgO(s)$$

$$2CuO(s) + C(s) \rightarrow 2Cu(s) + CO_2(g)$$

Suggest why, in industry, it is better to use carbon rather than magnesium to obtain copper from copper oxide.

Hint

Consider both the expense involved and also the ease with which the metal required can be separated from the mixture of products.

2 The reaction between aluminium and iron oxide is known as the thermite reaction. The diagram shows how the thermite reaction can be carried out.

The magnesium ribbon is lit to ignite the reaction mixture. The reaction between aluminium and iron oxide is very exothermic. The word equation for the reaction is:

   aluminium + iron oxide → aluminium oxide + iron

   a) What does the reaction suggest about the reactivity of aluminium compared with the reactivity of iron? Explain your answer.

   b) Explain which element is oxidised in this reaction.

   c) The thermite reaction can be used to join together two rails on a railway line.

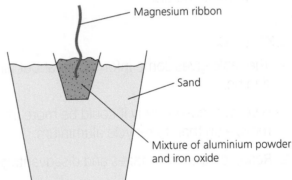

Magnesium ribbon

Sand

Mixture of aluminium powder and iron oxide

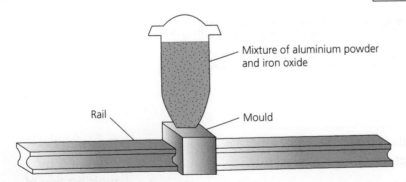

Mixture of aluminium powder and iron oxide

Rail

Mould

The reaction mixture is ignited and molten iron pours into the mould. The iron solidifies to create a join between the two rails. Explain why the iron produced in the reaction is molten.

Extend

1 Iron is obtained from the ore haematite, which contains iron oxide.
Titanium is obtained from the ore rutile, which contains titanium oxide.

Iron

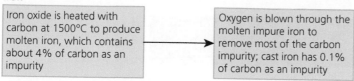

| Iron oxide is heated with carbon at 1500°C to produce molten iron, which contains about 4% of carbon as an impurity | → | Oxygen is blown through the molten impure iron to remove most of the carbon impurity; cast iron has 0.1% of carbon as an impurity |

Titanium

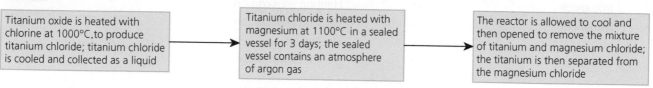

| Titanium oxide is heated with chlorine at 1000°C to produce titanium chloride; titanium chloride is cooled and collected as a liquid | → | Titanium chloride is heated with magnesium at 1100°C in a sealed vessel for 3 days; the sealed vessel contains an atmosphere of argon gas | → | The reactor is allowed to cool and then opened to remove the mixture of titanium and magnesium chloride; the titanium is then separated from the magnesium chloride |

a) Explain why the production of low-carbon steel uses oxygen, but the production of titanium uses argon and not oxygen.

b) There is less titanium than iron in the Earth's crust. Other than the scarcity of titanium, suggest reasons why titanium costs more than iron.

c) Two reactions that occur in the production of iron and titanium are:

$$2Fe_2O_3 + 3C \rightarrow 4Fe + 3CO_2$$

$$TiCl_4 + 2Mg \rightarrow 2MgCl_2 + Ti$$

Titanium can be used to produce iron from iron oxide. The equation for the reaction is:

$$2Fe_2O_3 + 3Ti \rightarrow 4Fe + 3TiO_2$$

Use these equations to obtain the relative reactivities of iron, magnesium and titanium. Give your reasons.

» Extracting metals using electrolysis

Worked example

Sodium is made by electrolysing sodium chloride. Write a word equation to represent this electrolysis.

sodium chloride → sodium + chlorine

Know

1 What is meant by the term electrolysis?

2 a) What is the name of the most common ore of aluminium?

b) Give the name of the aluminium compound that is present in this ore.

Apply

1 The table gives some information about the temperature required to produce a metal by heating its oxide with carbon. It also gives the most cost-effective method of extraction.

Metal oxide	Minimum temperature to produce the metal by heating with carbon (°C)	Most cost-effective method of extraction of the metal
Aluminium oxide	2027	Electrolysis
Calcium oxide	2170	Electrolysis
Copper oxide	105	Heating with carbon
Iron oxide	747	Heating with carbon
lead oxide	350	Heating with carbon
Magnesium oxide	1625	Electrolysis
Zinc oxide	930	Heating with carbon

The order of reactivity of the metals listed in the above table is:

calcium > magnesium > aluminium > zinc > iron > lead > copper

Use the information in the table to explain how the method of extraction of a metal is related to its reactivity and the energy involved.

2 The table shows the results of electrolysing three molten metal halides (i.e. compounds of a metal and a halogen).

Molten compound	Observation at negative electrode	Substance formed at negative electrode	Observation at positive electrode	Substance formed at positive electrode
Sodium chloride	Flashes of light	Sodium	Pale green gas	Chlorine
Lead iodide	Shiny liquid ball		Purple gas	
Compound X	Shiny liquid ball		Orange–brown gas	

a) Identify the elements formed at the negative and positive electrodes when lead iodide is electrolysed.

b) Suggest a possible identity for compound X.

c) Describe what you might expect to see at the negative and positive electrodes when molten potassium iodide is electrolysed.

Extend

1 Aluminium is produced in industry by the electrolysis of a solution of aluminium oxide dissolved in molten cryolite at a temperature of around 930°C using carbon electrodes. Aluminium is produced at the negative electrode and oxygen is produced at the positive electrode. The positive electrodes gradually get smaller and smaller during the electrolysis and eventually have to be replaced.

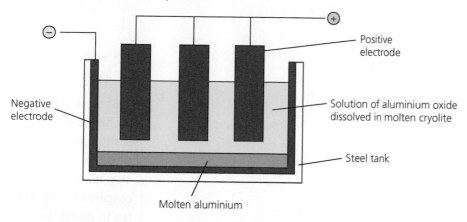

Negative electrode

Positive electrode

Solution of aluminium oxide dissolved in molten cryolite

Steel tank

Molten aluminium

a) Give the formula of aluminium oxide.

b) Why is the aluminium oxide dissolved in molten cryolite rather than using molten aluminium oxide?

c) Suggest why the positive electrodes get smaller and smaller during the electrolysis.

d) Iron could be produced by electrolysing molten iron oxide. However, it is not produced this way. Instead it is produced by heating iron oxide with carbon. Explain why.

e) Why is aluminium not produced by heating aluminium oxide with carbon?

f) Iron, in the form of mild steel, is used to make car bodies. Some car bodies are made of aluminium.

   i) What property of steel makes it more suitable than aluminium for making car bodies?

   ii) State two properties of aluminium that make it a better metal than iron for making car bodies.

15 Breathing

» The respiratory system

Worked example

Explain the difference between respiration and breathing.

Respiration is the chemical reaction that takes place in every cell, in which glucose is broken down to release energy.
Breathing is the movement of air into and out of our lungs.

Know

1 Fill in the missing words:

The respiratory system is made up of a tube that runs from the mouth called the _____. This is lined with tiny hairs called _____ and is protected by rings of _____.

The two bronchi split into many smaller tubes called _____. These end in small air sacs called _____.

2 What are the two structures labelled A and B on the diagram?

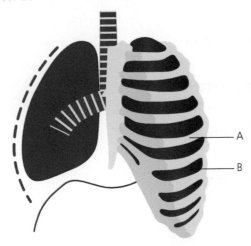

Apply

1 Name the rings labelled C on the diagram. Describe the function of these rings.

2 The substances in cigarette smoke paralyse and kill the ciliated cells of the trachea and bronchi. Explain how this could lead to:

   a) a cough

   b) more chance of lung infections.

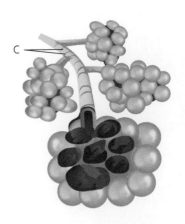

Extend

1 The alveoli are surrounded by a network of fine capillaries:

a) On a copy of the diagram draw arrows to indicate the movement of:

i) oxygen

ii) carbon dioxide.

> **Hint**
>
> Remember to get these the right way round – carbon dioxide is produced as a waste product and so leaves the blood, while oxygen is needed for aerobic respiration and so enters the blood.

b) The air that we breathe in and out is composed of around 80% nitrogen. Find out what happens to the nitrogen in the air that we breathe into our lungs.

2 Emphysema is a condition in which the walls separating the alveoli are broken down:

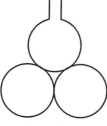

Normal lung

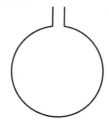

Lung with emphysema

> **Hint**
>
> Don't forget that if you measure the diameter of a circle, you then need to divide by 2 to get the radius, r.

a) The circumference of a circle is given by $2\pi r$, where r is the radius of a circle and $\pi = 3.14$.

Calculate the circumference of the alveoli in the diagram of:

i) a normal lung

ii) a lung with emphysema.

b) How many times bigger is the circumference of the alveoli of the normal lung compared with the lung with emphysema?

c) Predict what effect this will have on the gas exchange in the alveoli.

» Alveoli

Worked example

The composition of air was analysed before and after it entered a person's lungs. Less oxygen was found in the air leaving the lungs compared with air entering the lungs. Explain this observation.

Oxygen diffuses from the air entering the lungs into the blood capillaries through the alveoli. The oxygen is used by the body for respiration.

Know

1 Copy and complete the table below, showing the composition of air in inspired and expired air, using the word 'More' or 'Less'.

	Inspired air (%)	Expired air
Oxygen	20	Less
Carbon dioxide	0.04	

2 The alveoli are specially adapted so that gases can move into or out of the blood. Describe how each of the following adaptations helps gas movement

a) large surface area

b) permeable

c) very thin walls

Apply

1 Copy and complete the following sentences by filling in the missing words:

The carbon dioxide in the blood moves out into the _____ by a process called _____. This is the movement of molecules from an area of _____ concentration to an area of _____ concentration.

2 Describe how the following causes problems when a person smokes cigarettes.

a) nicotine

b) tar

3 A smoker develops a bad cough and starts to find that exercise becomes more difficult. Explain the reason for each of the following:

a) a smoker's cough

b) a smoker's reduced ability to exercise

4 Describe how asthma affects how a person breathes.

Extend

1 The diagram shows a drawing of an alveolus in a human lung.

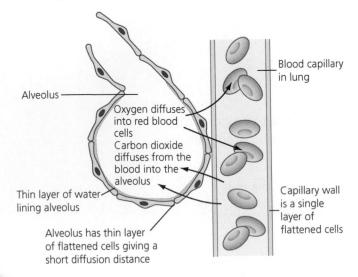

Blood capillary in lung

Alveolus

Oxygen diffuses into red blood cells

Carbon dioxide diffuses from the blood into the alveolus

Thin layer of water lining alveolus

Alveolus has thin layer of flattened cells giving a short diffusion distance

Capillary wall is a single layer of flattened cells

a) When we start to exercise our heart rate and breathing rate change very quickly. Use the diagram to explain the effect of each of the following on the events that take place in the alveoli:

   i) increased heart rate

   ii) increased breathing rate

b) Some people who smoke have damaged lungs and have problems breathing normally. Explain how each of the following will affect the movement of gases in the alveoli:

   i) The damaged cilia allow mucus to build up in the lungs.

   ii) The lungs have fewer elastic fibres.

» The process of breathing

Worked example

The volume of the chest cavity increases when we breathe in. Explain how the follow bring about this change in volume:

a) intercostal muscles
b) diaphragm

a) Intercostal muscles contract and move the rib cage outwards and upwards, increasing the volume of the chest cavity.
b) The diaphragm contracts and flattens, making more room in the chest cavity for the lungs.

Know

1 Copy and complete the following sentences by filling in the missing words:

   The process of breathing in and out is known as _____. The movement of air into the lungs is called _____ and the movement of air out of the lungs is called _____.

2 Identify structures D, E and F on the diagram on the right.

3 Put the following statements in the correct order for the events that occur when a person breathes in. Numbers 1 and 5 are done for you.

   1 intercostal muscles contract

      ribs move inwards and downwards/ribcage moves upwards and outwards

      diaphragm contracts/diaphragm relaxes

      diaphragm becomes curved/diaphragm flattens

   5 air moves into chest cavity

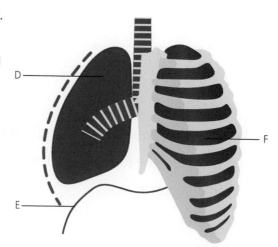

Apply

1 When we breathe in we use muscles to take air into the lungs. Breathing out does not normally require muscles to contract. Suggest why breathing out is normally a passive process.

2 When an athlete is competing in a race, their muscles are used to help them breathe out. Explain how this would help the athlete compete.

Hint

There are two sets of intercostal muscles – internal and external. Carry out some research to find out how these differ in their function.

Extend

1 The table below shows the lung volumes for three different people.

Person	Age	Air passing into the lungs (cm³)	
		At rest	During exercise
A	25	500	5000
B	25	500	4000
C	65	400	3500

Person A is a trained athlete whereas B and C are not.

a) Explain how more air is taken into the lungs during exercise.

b) Person A is an athlete, whilst person B is the same age but not an athlete.

i) Explain why the volume of air at rest is the same for the athlete and the non-athlete.

ii) Explain why the volume of air during exercise for the athlete is much higher than for the non-athlete.

c) Use the data to describe how age affects the function of the lungs.

d) Suggest one other explanation for the difference in results seen for A, B and C.

2 The diagram shows the plan for a model of the breathing system:

a) Describe how each of the following is represented in the model:

i) trachea

ii) lungs

iii) diaphragm

iv) rib cage

v) bronchi

b) Discuss how well each of the above is represented in this model.

c) Explain why the following structures are not represented:

i) intercostal muscles

ii) alveoli

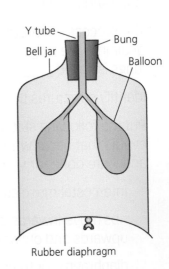

Y tube
Bung
Bell jar
Balloon
Rubber diaphragm

16 Digestive system

» Food groups and a balanced diet

Worked example

Describe the benefits of a person eating the following foods as part of a balanced diet:

a) carbohydrates   b) proteins   c) lipids

a) Carbohydrates are a good source of energy and can be broken down to allow activities such as muscle contraction to take place.

b) Proteins are a source of the building blocks required for growth and repair of body tissues.

c) Lipids such as fats and oils are very high in energy. Small amounts of lipids are required for a healthy diet.

Know

1 Copy and complete the following sentences by filling in the missing words:

After visiting a doctor a person was told to cut down on saturated _____ in their diet. This could be done by reducing foods such as _____ and _____. Continuing to eat a large proportion of these foods could lead to an increase in _____ levels in the blood and possibly lead to the development of _____ disease.

2 Lipids are classified as either fats or oils.

a) Describe the difference between fats and oils.

b) Name one fat and one oil.

3 A balanced diet includes the correct amounts of different vitamins. Describe the function of the following vitamins:

a) vitamin A   b) vitamin C   c) vitamin D

Apply

4 A young person's diet for a day is shown in the table.

a) Which of components of the person's diet contain high levels of:

i) protein   ii) carbohydrate   iii) fat

Breakfast	Lunch	Dinner
Cereal	Salmon	Rice
Milk	Chips	Chicken
Buttered toast	Peas	Eggs
Tea	Water	Coffee

b) If the person wanted to change the meals to a vegetarian diet, what could they substitute for the foods high in protein?

Extend

1 Sucralose is an artificial sweetener that is around 1000 times sweeter than normal sugar (sucrose). Sucralose was invented in 1976 and in 2004 it was allowed to be used in a wide variety of food products in Europe.

a) Look at the diagram and describe how sucralose is different from sucrose.

b) Suggest why it took almost 30 years before sucralose was allowed to be used in food.

c) Sucralose is unable to be broken down in the digestive system and so contributes no calories to a person's diet.

   i) Describe how most food is used by the body after it is eaten.

   ii) Use your knowledge of enzymes to suggest why sucralose is unable to be broken down.

2 A company released a new breakfast biscuit. It claims to offer everything required for a healthy, balanced breakfast meal. The table below shows the nutritional information for the product.

	One biscuit (30 g) contains:	% recommended for a healthy breakfast
Calories	100	25
Fat...	1.3 g	2
...of which saturated	0.3 g	2
Carbohydrates...	27 g	–
...of which sugars	24 g	30
Fibre	0.7 g	–
Protein	1 g	2
Salt	2 g	30
Calcium	0	0
Iron	1 mg	10
Vitamin A	0	0
Vitamin C	0	0
Vitamin D	1.5 µg	30

a) How many biscuits would be required to meet 100% of the energy needs of the child?

b) A person eats enough biscuits to meet 100% of their energy needs for breakfast. Calculate the percentage of the recommended amount for each of the following that the person gets.

   i) fat   iii) protein

   ii) sugars   iv) salt

c) The breakfast bar claims to be 'full of vitamins and minerals'. To what extent is this true?

d) Suggest what effect on health the following could have.

   i) high salt   iii) low protein   v) low vitamin C

   ii) high sugar   iv) low calcium

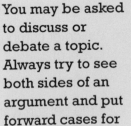

» The digestive system

Worked example

When eating food it is first taken into the mouth to start the process of digestion. Describe what the teeth and tongue do to the food in the mouth.

The food is crushed and broken down using the teeth and tongue. This breaks the food into smaller pieces, giving it a much larger surface area. This means that digestion takes place faster and the food can be swallowed easily.

Know

1 Saliva is secreted from glands in the mouth and helps with the process of digestion. Describe two functions of saliva in the mouth.

2 Using the key words in the box and your knowledge of enzymes, copy and complete the following sentences.

> carbohydrates   increase   faster   pancreas   proteins

Enzymes _____ the rate of reaction, allowing digestion of food to happen _____. In the mouth, saliva is mixed with the food and contains enzymes that start the breakdown of _____. The stomach produces another enzyme that breaks down _____. Many of the enzymes used in the small intestine are produced in the _____.

3 Look at the diagram on the right and name the structures labelled A, B, C and D.

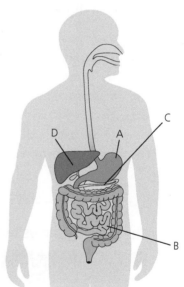

Apply

1 Short bowel syndrome (SBS) is a condition in which a lot of the small intestine is missing.

   a) What is the main purpose of the small intestine?

   b) How might the health of a person with SBS be affected by the condition?

   c) Suggest how the diet of the person could be modified to help with the problems associated with SBS.

Extend

1 Coeliac disease is a condition in which a person reacts badly to gluten in the diet. Gluten is a protein found in several carbohydrate-rich foods.

   a) Name two foods that could contain gluten.

   b) Explain how a person with coeliac disease would have to modify their diet.

> **Hint**
>
> A question that asks you to 'suggest' an answer is usually very open-ended. Any sensible suggestion that answers the question will be accepted.

c) The gluten causes irritation of the intestines and may lead to the finger-like projections (villi) on the inside of the intestine being shorter. Explain what effect smaller villi will have on the absorption of food in the small intestine.

d) Explain why a person with coeliac disease might suffer from the following:

   i) diarrhoea

   ii) low weight

   iii) anaemia

Hint !

Remember to answer this type of question by mentioning surface area. How would the surface area change if villi became shorter, and how does surface area affect absorption of food?

» The digestive process

Worked example

The complete digestion of our food requires both mechanical and chemical digestion. What is meant by:

a) mechanical digestion
b) chemical digestion?

a) Mechanical digestion is the breakdown of food using the teeth and tongue in the mouth. This breaks apart large pieces of food into smaller pieces.
b) Chemical digestion occurs in the mouth, stomach and small intestine, and involves the use of chemicals such as enzymes. Enzymes break down large food molecules into smaller molecules that can be absorbed into the blood.

Know

1 The drawing shows two types of teeth in the mouth.

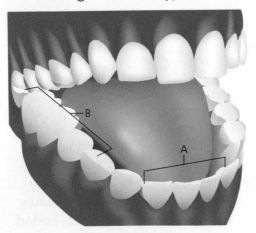

a) Identify the teeth labelled A and B.

b) Describe the function of:

   i) A

   ii) B

2 A person eats a cheese-and-ham pizza. Describe the roles of each of the following in the digestion of the pizza:

   a) carbohydrase

   b) protease

   c) lipase

Apply

1 A student set up three test tubes to investigate the rate of reaction of the enzyme lipase. Lipase breaks down the fats in our food. The three test tubes contained the following:

   • test tube 1: fat + lipase

   • test tube 2: fat + lipase + bile

   • test tube 3: fat + water

   The table shows the results.

Test tube	Contents	Rate of reaction (g/min)
1	fat + lipase	3
2	fat + lipase + bile	24
3	fat + water	0

   a) Explain why test tube 2 has a much faster rate of reaction that test tube 1.

   b) Suggest what test tube 3 tells us about the enzyme lipase.

Extend

1 Bacteria are found living in the digestive system of all animals. They help with the digestion of the food that passes through the gut.

   a) What do bacteria produce that allows them to break down the food we eat?

   b) Suggest why bacteria are able to digest some food that we cannot break down.

   c) Explain one possible benefit of this relationship to:

   i) the bacteria

   ii) the animal host.

   d) Animals are not born with any bacteria in the digestive system. Suggest how young animals acquire bacteria in their guts.

   e) Sometimes, the bacteria living in the gut are destroyed or removed (for example due to the use of antibiotics). This may mean that other bacteria can grow instead, causing the person to become ill. What could be another symptom of reduced bacteria in the gut?

   f) Suggest how the natural gut bacteria prevent the growth of the harmful bacteria.

   g) How might harmful bacteria cause a person to become ill?

17 Respiration

» Aerobic respiration

Worked example

Aerobic respiration involves the breakdown of glucose using oxygen. Describe where each of the following comes from:

a) oxygen
b) glucose

a) When we breathe, air moves into our lungs. The oxygen in the air diffuses into our blood and is taken to our cells for aerobic respiration.

b) A lot of the food we eat is digested and breaks down into glucose in our small intestine. The glucose is then absorbed into our blood and is transported to our cells for respiration.

Know

1 Copy and complete the equation for respiration shown:

$$\text{glucose} + \text{oxygen} \rightarrow \underline{\hspace{2cm}} \ \underline{\hspace{2cm}} + \underline{\hspace{2cm}}$$

2 Aerobic respiration releases energy from the glucose molecules. State three cell processes that rely on this release of energy.

Apply

1 A study was carried out in which the numbers of mitochondria in different types of cell were counted. The results are shown in the table:

Type of cell	Number of mitochondria
A	850
B	10
C	900

Cell type A was a muscle cell.

a) Suggest the type of cell for:

i) cell B

ii) cell C.

b) It was observed that the number of mitochondria in cell A increased after the person exercised for several months. Explain this observation.

> **Hint**
>
> Think carefully about all of the processes in cells that you have studied that use energy. These may be from other topics. Bringing together information from different areas of the subject is an important skill.

Extend

1 A plant was kept in a sealed container with 0.1% carbon dioxide and 20% oxygen. It was then exposed to light for 12 hours followed by darkness for 12 hours. The concentrations of oxygen and carbon dioxide in the container were measured. The results were plotted on a graph:

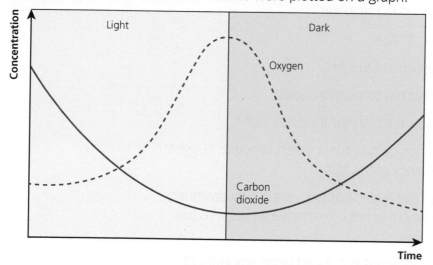

a) Use the graph and your knowledge to explain the following:

i) the rise in oxygen in the light

ii) the fall in carbon dioxide in the light

iii) the fall in oxygen in the dark

iv) the rise in carbon dioxide in the dark

b) The experiment was set up with 0.1% carbon dioxide in the container, but there was only 0.07% at the end of the experiment. Suggest why the concentration of carbon dioxide after 24 hours is less than at the start.

c) How might temperature affect the results?

d) In a further experiment the intensity of the light was investigated. Predict and explain the effect of increasing light intensity on the production of oxygen.

2 The diagram shows the gills of a fish.

> **Hint**
>
> When answering questions that involve temperature, always think about enzymes. Enzymes are sensitive to temperature and control the rate of many reactions.

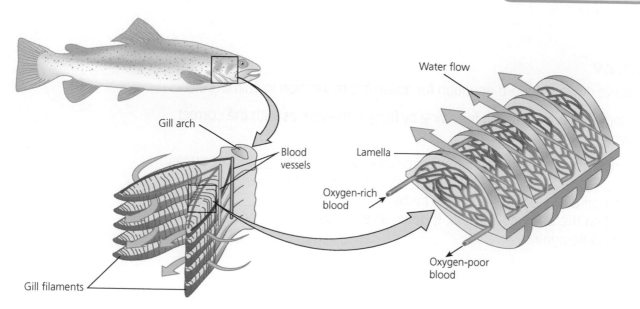

a) The gills are used to exchange gases with the water surrounding the fish.

   i) Which gas would be removed from the water?

   ii) Why does the fish need this gas?

   iii) Which gas would be returned to the water?

   iv) Where does this gas come from?

b) Blood flows through blood vessels in the gills.

   i) Name the blood vessels that run through the gills.

   ii) How are these blood vessels adapted for their function?

c) The gills are made up of a series of structures called lamellae. Suggest how these increase the efficiency of the gills.

d) Fish keep water moving over the surface of the lamellae by swimming or by gulping water. Explain how this helps exchange gases with the blood.

e) Fish that are kept in acidic conditions produced large amounts of thick mucus that cover the gills. Describe how this might affect the fish.

» Anaerobic respiration in humans

Worked example

The cells in our bodies respire aerobically for most of the time. Sometimes, however, they respire anaerobically. Explain why a cell might respire anaerobically.

A cell may have to respire anaerobically instead of aerobically because it could be very active and using up the available oxygen very quickly. Also, in some situations the oxygen supply to the cell might be blocked, forcing cells to respire anaerobically for a short time.

Know

1 Write down the word equation for anaerobic respiration in animal cells.

2 Copy and complete the following by filing in the spaces with the correct word:

Anaerobic respiration means respiration without _____. It produces _____ _____ because _____ is not fully broken down into carbon dioxide and _____. The build-up of _____ _____ is toxic and so the body cells break it down as soon as _____ becomes available again.

3 The energy released by anaerobic and aerobic respiration was measured and recorded in a table:

Type of respiration	Energy released (kJ)
Aerobic	200
Anaerobic	10

a) What is meant by

   i) aerobic respiration

   ii) anaerobic respiration?

b) Calculate the energy released by anaerobic respiration as a proportion of the energy released by aerobic respiration.

c) Suggest why there is a difference between the energy released by the two processes.

Apply

1 An 800 m runner competes in a race. During the race his breathing rate increases.

   a) Explain why the breathing rate increases during the race.

   b) Explain why an athlete continues to breathe heavily even after the race has finished.

Extend

1 The graph shows the uptake of oxygen at rest and when a person exercises.

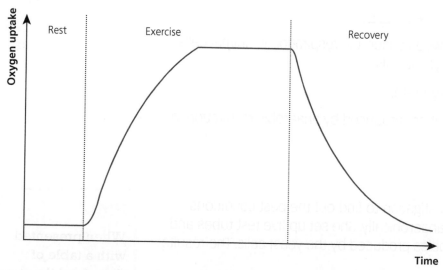

   a) Oxygen is used up at rest and also during exercise.

      i) Name the process that uses up oxygen.

      ii) Where does this process take place?

      iii) Name some of the body processes that require oxygen.

   b) Explain the following:

      i) There is an increase in oxygen uptake when the person starts to exercise.

      ii) The oxygen uptake levels off.

c) When a person starts to exercise there will be changes in the body.

  i) Describe the changes in the breathing system that causes an increase in oxygen uptake.

  ii) Describe the changes in the circulatory system that accompany an increase in oxygen uptake.

d) Explain the shape of the curve for oxygen uptake during the recovery period.

Hint

This question asks for a description of any changes. It does not want you to go on to explain a reason for those differences. Remember to read the stem of any question carefully.

» Anaerobic respiration in microorganisms

Worked example

Anaerobic respiration can take place in animal cells and also in many microorganisms such as yeast. In yeast we call it fermentation.

Describe the differences between anaerobic respiration in animal cells and anaerobic respiration (fermentation) in yeast cells.

Anaerobic respiration in animal cells involves the production of lactic acid, whereas in yeast it produces the alcohol ethanol. Fermentation in yeast also produces the gas carbon dioxide.

Know

1 Copy and complete the word equation for anaerobic respiration in yeast cells:

glucose → _____ + _____

2 Describe two similarities between anaerobic respiration in animal cells and anaerobic respiration in yeast cells.

3 Name three types of microorganism.

4 Name three types of food that are produced by anaerobic respiration in microorganisms.

Apply

1 A student carried out an investigation to find out the best conditions needed for yeast to respire anaerobically. She set up five test tubes and recorded the number of bubbles produced by the yeast cells. The results are shown in the table.

Test tube	Yeast (cm³)	Glucose (cm³)	Temperature	Number of bubbles
1	5	5	10	5
2	5	5	20	43
3	5	5	50	12
4	5	0	20	0

a) What gas did the bubbles contain?

b) Explain the results for test tubes 1, 2 and 3.

c) What does test tube 4 tell us about yeast cells?

Hint

When presented with a table of data, read the stem of the question carefully. You may be asked to simply describe the trends in the data or, as in this case, to explain the results that were obtained.

2 Explain what happens to bread when it is baked at high temperatures in an oven.

Extend

1 The diagram shows an experiment to investigate the activity of yeast, a microorganism.

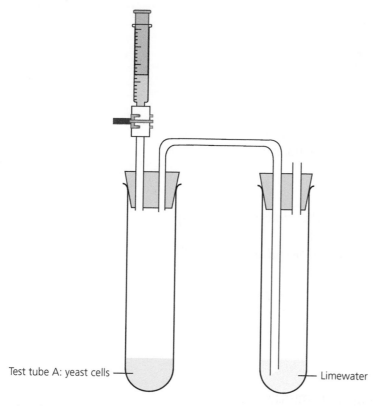

Test tube A: yeast cells —   — Limewater

a) After several minutes, bubbles were observed being produced from the end of the tube in the limewater. The apparatus was set up with no oxygen.

　i) Name the process taking place in the yeast cells that produces bubbles of gas.

　ii) Write an equation for this process.

b) Name the gas produced.

c) Describe the change that will occur to the limewater in the test tube.

d) What needs to be added to the yeast cells to allow them to carry out this process?

e) Name the substance produced by the yeast cells in test tube A.

f) The number of bubbles being produced could be counted. Explain how temperature would affect the rate at which bubbles are produced.

18 Photosynthesis

» The process of photosynthesis

Worked example

Photosynthesis produces glucose molecules that can be used for different functions in a growing plant. Describe three of the functions that glucose can be used for.

Glucose can be used for:
- respiration, where it is combined with oxygen to release the energy that keeps plant cells alive
- the production of cellulose, which is used to make the cell walls of new plant cells
- the production of starch, which is an insoluble storage molecule, allowing the glucose to be used later as the plant grows.

Know

1 Copy and complete the paragraph by filling in the missing words.

   Plants carry out photosynthesis by using energy from the _____ to combine _____ _____ with water to make _____ and glucose. The _____ of the plant absorb the water while the _____ absorb the _____ _____.

2 What is meant by a limiting factor?

Apply

1 Without plants, life would be very different. They are essential for organisms to survive on Earth. Give examples of how plants contribute to the following:

   a) allowing living things to respire

   b) production of food

   c) production of medicines

   d) production of materials for clothing

2 A student did an investigation in which he measured the rate of photosynthesis in a plant kept under different conditions. The results are shown in the table on the next page.

Plant	Temperature (°C)	Carbon dioxide concentration (%)	Rate of photosynthesis
A	20	1	100
B	20	2	100
C	25	1	120
D	25	2	120

a) What is the limiting factor in the investigation? Explain your answer.

b) Suggest how this factor could limit the rate of photosynthesis.

c) Apart from carbon dioxide and temperature, name another factor that might be important in controlling the rate of photosynthesis in this plant.

Extend

1 The rate at which a plant photosynthesises can be investigated using the apparatus shown in the diagram. The number of bubbles of gas produced by the plant can be counted.

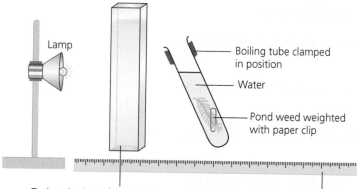

Lamp

Boiling tube clamped in position

Water

Pond weed weighted with paper clip

Tank or beaker of water between pond weed and lamp   Metre ruler

a) The plant produces bubbles of gas.

   i) What gas do the bubbles contain?

   ii) How could you show that the bubbles are made from this gas?

b) Write the equation for photosynthesis.

c) Suggest why a beaker of water is placed between the light and the plant.

d) The light source can be moved away from the plant.

   i) What effect will this have on the light intensity?

   ii) What effect will this have on the number of bubbles produced?

e) A green filter is placed between the light and the plant so that the plant only receives green light. Suggest what effect this would have on the rate at which the plant produces bubbles.

Hint

Think about the colour of the leaves of plants. To be able to see a green leaf, the leaf must be reflecting the green light. If the light is being reflected, could it be used for photosynthesis by the plant?

2 The diagram shows two simple food chains, one from woodland and one from a deep-sea hydrothermal vent. Hydrothermal vents are deep under the ocean where no light reaches.

Woodland food chain:

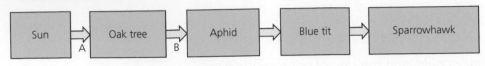

Hydrothermal vent food chain:

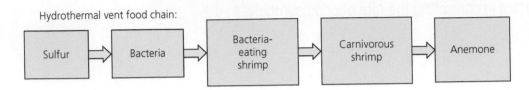

a) Name the process indicated by arrow A.

b) Describe what is happening at arrow B.

c) Apart from the different organisms involved, describe:

   i) one difference between the woodland and hydrothermal food chains

   ii) two similarities between the woodland and hydrothermal food chains.

d) Approximately 10% of the energy is lost between each of the stages of the food chains.

   i) If 10 000 kJ was available to the oak tree, how much would be available to the sparrowhawk?

   ii) What process releases energy at each of the stages of a food chain?

   iii) Suggest how some of this energy is lost between stages.

3 The graph shows the production or uptake of oxygen by a plant as the light intensity increases:

a) Explain why:

   i) oxygen is removed at low light intensity

   ii) oxygen is produced at high light intensity

   iii) the curve starts to level off.

b) The point at which the line crosses the x-axis is called the compensation point.

   i) How much oxygen is being produced or removed at the compensation point?

   ii) Explain your answer.

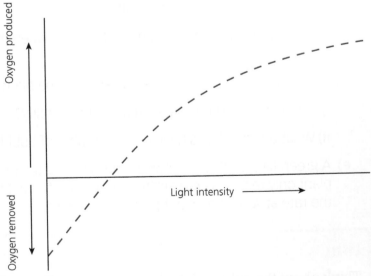

» Leaf structure

Worked example

Describe the functions of each of the following leaf components:

a) palisade layer
b) spongy layer
c) waxy cuticle

a) The palisade layer of the leaf is made up of cells that are packed full of chloroplasts. These carry out photosynthesis.
b) The spongy layer has cells with air spaces between them. These air spaces allow gases to be exchanged and water to evaporate. The cells also contain chloroplasts and help with photosynthesis.
c) The cuticle is a waxy layer that prevents water from evaporating from the outer leaf cells. This prevents the leaf from drying out.

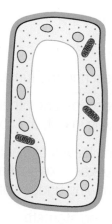

Know

1 Copy the diagram of the plant cell and add a label to show where photosynthesis takes place.

2 Describe how each of the following helps a leaf carry out its function:

  a) large surface area

  b) very thin

  c) has a network of veins

3 Copy the diagram and add labels to show the following:

  a) waxy cuticle

  b) spongy layer

  c) air spaces

  d) palisade layer

  e) stomata

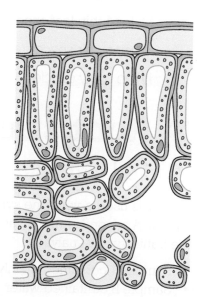

Apply

1 A student measured the leaf area of three leaves from three different types of plant. She recorded them in a table:

Plant	Leaf 1 area (cm²)	Leaf 2 area (cm²)	Leaf 3 area (cm²)	Average leaf area (cm²)
A	40	41	39	40
B	3	1	2	2
C	15	15	12	14

  a) Which leaf would be best for absorbing as much sunlight as possible?

  b) Which leaf comes from a plant growing in a hot, dry environment?

  c) Suggest one other feature of the leaves that the student could have measured.

Extend

1 Leaves were taken from two different plants, A and B. The leaves were then cut and studied under a microscope:

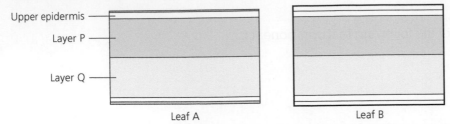

Leaf A   Leaf B

a) Which of the two leaves is from a plant adapted to growing in a hot, dry environment? Explain your answer.

b) Name the following:

   i) Layer P

   ii) Layer Q

c) Describe the function of the waxy cuticle.

d) Describe one other feature that would show that a leaf is adapted to living in hot, dry conditions.

e) The leaves of plants growing in water also show many adaptations. Suggest explanations for the following adaptations shown by plants growing under water.

   i) no stomata

   ii) long thin leaf shape

   iii) many air spaces (aerenchyma)

   iv) no waxy cuticle

» Root structure

Worked example

Scientists used a very fine syringe to take a sample of liquid from the stem of a plant. A second sample was taken from a different position on the stem and the liquids were analysed.

- Sample A contained mainly water with some minerals.
- Sample B contained water with sugar dissolved in it.

Which sample was taken from the xylem vessels and which was taken from the phloem? Explain your answer.

Sample A was taken from the xylem as this transports water and minerals from the soil up to the leaves of the plant.

Sample B was taken from the phloem as this transports the sugars made by photosynthesis to the rest of the plant.

Know

1 Explain how the roots of a plant can help with the following:

   a) preventing the plant from blowing over in strong winds

   b) allowing the plant to photosynthesise

2 Xylem vessels and phloem tissue are both used for transport in a plant.

   a) How is the xylem adapted for its function?

   b) Describe two ways in which xylem and phloem differ in their structure and function.

3 Use the diagram to explain the role of the root hair cells and describe how they are adapted for their function.

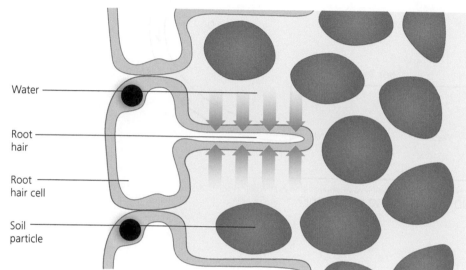

Water

Root hair

Root hair cell

Soil particle

Apply

1 A new fertiliser is being marketed that promises to allow gardeners to grow better plants. Two of the key ingredients are:

   • magnesium sulfate

   • ammonium nitrate.

   Explain how each of these ingredients could help to improve the growth of the gardener's plants.

Extend

1 Plants transport water and minerals from the roots to the leaves. They then transport water and sugars to the other tissues of the plant. This process has been modelled as shown in the diagram below.

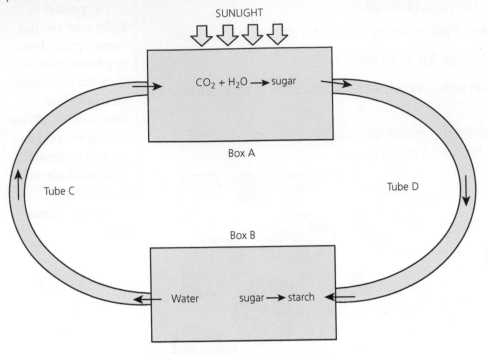

SUNLIGHT

$CO_2 + H_2O \longrightarrow$ sugar

Box A

Tube C

Tube D

Box B

Water

sugar $\longrightarrow$ starch

a) In this model, what do the following represent?

 i) Box A

 ii) Box B

 iii) Tube C

 iv) Tube D

b) How is sugar produced in Box A?

c) i) Why does water travel from Box B to Box A?

 ii) Name the process by which the water moves.

d) Suggest what causes the sugar solution to move from Box A to Box B.

e) Give one property of starch that makes it a good storage form of sugar.

f) Plants are known to transport sugars day and night. Discuss why this model may not give a full explanation of transport in plants.

Hint

Many processes in biology can be modelled, but not all models are completely correct. An important skill is to be able to understand how models represent the process, but also to understand their faults.

19 Evolution

» The theory of evolution by natural selection

Worked example

The last universal common ancestor was likely to have been a single-celled organism that existed on Earth some 3.5 billion years ago.

Why is it called the last universal common ancestor?

The single-celled organism is the life form from which all other life on Earth developed. It is the point at which life began to change into the different species that we see around us today.

Know

1 Using the key words in the box, copy and complete the following sentences.

species	DNA	biodiversity	population

Variation is the difference between individuals due to their _____. The difference between individuals within a _____ is known as _____. It can also mean the different types of _____ found in different habitats.

2 A scientist has stated that 'evolution cannot take place if there is no variation'. Explain why evolution requires variation.

Apply

1 The following statements describe the stages that occur during the process of evolution. Put these statements into the correct order.

- The advantageous characteristics are passed on to the next generation.

- Some individuals are better adapted than others.

- Individuals within a population show variation.

- These are therefore more likely to reproduce.

- These individuals are more likely to survive to a mature age.

2 Around 66 million years ago the dinosaurs became extinct. Use your knowledge of the theory of evolution to explain how a species could become extinct.

Extend

1 The diagram below shows the relationship between five species of mammal:

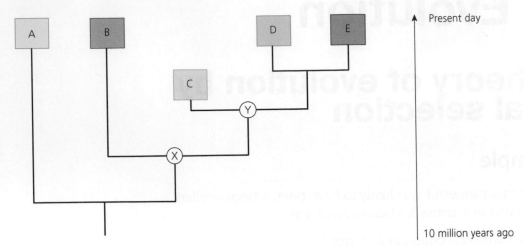

a) Identify on the diagram the point at which the last universal common ancestor existed.

b) Identify the common ancestor of species C, D and E.

c) Identify the two species of mammal that are most closely related.

d) Explain what has happened to species C.

e) The formation of new species is known as speciation. Describe the conditions required for speciation to take place.

» Charles Darwin

Worked example

In 1859 Charles Darwin published his theory of evolution. Describe Darwin's theory of evolution.

Darwin said that all organisms show variation. There may be competition between the organisms and so some may have an advantage. These ones are most likely to survive and most likely to reproduce. The favourable characteristics of the survivors will be passed on to the next generation. Over a long period of time this will cause a change in the organism.

Know

1 Explain what Charles Darwin meant by the term 'natural selection'.

2 Darwin travelled to the Galapagos Islands aboard the survey ship the *Beagle*.

a) What did Darwin notice on the voyage of the *Beagle* that helped him develop his theory when he returned back to England?

b) What made the Galapagos Islands an ideal place for Darwin to make observations supporting the theory of evolution?

Apply

1 In 1864 Herbert Spencer first used the term 'survival of the fittest' to describe the natural selection that Darwin described in his book.

a) What is meant by 'survival of the fittest'?

b) Suggest why 'survival of the fittest' is not always an accurate description of the process of evolution.

2 The drawing shows some of the birds that Darwin sent back from his visit to the Galapagos Islands.

It is thought that the birds all had a common ancestor with a single beak shape. Suggest how the birds could have developed into the variety of beak shapes seen.

> **Hint** !
>
> Think carefully about what we mean by being the 'fittest'. It often means being the biggest, strongest or most healthy. Sometimes, however, other features (such as small size) may be an advantage.

Extend

1 In the mid-nineteenth century Alfred Russel Wallace travelled extensively around the islands of Southeast Asia, observing and making notes on the variation between the species of animals and plants on the islands he visited.

In 1858 he sent details of the theory he had developed to Charles Darwin. This prompted Darwin to publish his own theory.

a) Why would there be differences in species between islands in the same area?

b) What was the theory that both Wallace and Darwin came up with?

c) What is meant by the term 'species'?

d) Suggest why Darwin was prompted to publish his own theory after receiving the letter from Wallace.

e) Darwin had worked on his own theory for many years before publishing. Why might a scientist continue to research a theory before making it public?

South East Asia

Mindanao

Halmahera

Sumatra   Borneo   Ceram   Tanimbar Island

Sulawesi

Java   Buru

Bali   Timor   Babar

Wallace line   Lombok

Australia

Key
☐ Present-day mainlands
☐ Mainland extensions by lower sea level during the ice age

> **Hint** !
>
> Think about the way scientists work. One of the most important aspects of research is providing sufficient evidence to support a theory. This is also true for Darwin's theory of evolution.

2 The diagram shows the front limb of four animals with backbones – human, dog, bird and whale. The shaded bones are known to have the same origins.

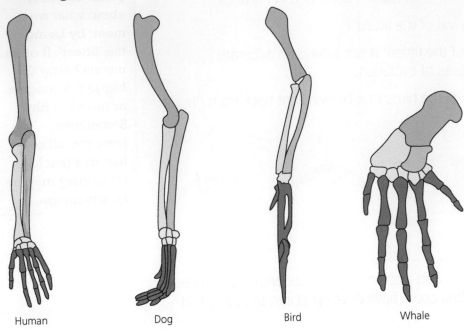

Human   Dog   Bird   Whale

a) Describe the different roles of the limb in the following animals:

i) dog

ii) bird

ii) whale.

b) Suggest what 'from the same origin' means.

c) Use Darwin's theory of evolution to explain how the limb of the bird could have developed.

d In what way does the structure of the four limbs support the theory of evolution?

» Biodiversity

Worked example

It is important that we try to protect and maintain the biodiversity in the world around us.

a) Give two reasons why maintaining biodiversity is a benefit.

b) Give examples of the ways in which humans have tried to maintain biodiversity.

a) Maintaining biodiversity is important because:
- species are less likely to become extinct in a diverse habitat
- many species may be of potential benefit to us, such as in the production of new medicines.

b) We have tried to preserve biodiversity by setting up national parks, conservation zones and introducing laws to protect the environment.

Know

1 Give examples of each of the following types of biodiversity.

   a) biodiversity within a species

   b) biodiversity between species

   c) biodiversity between habitats

Apply

1 The numbers of different species of mammal found in three different habitats in Argentina were measured and the results recorded in a table:

Habitat	Number of mammal species
Mountain	45
Forest	100
Grassland	120

   a) Suggest how the number of species in a habitat could be measured.

   b) Explain which habitat has the greatest biodiversity.

   c) Suggest why this study may not give a true value for the biodiversity of each habitat.

Extend

1 Humans can have a huge impact on the biodiversity of the world around them. One area of human influence is the mining of raw materials to produce the goods we require. Mining can involve deforestation of vast areas of forest.

   a) i) Suggest why activities such as mining require the cutting down of trees.

   ii) Why else might trees be cut down?

   b) Explain how each of the following could influence the animals and plants as a result of deforestation:

   i) increased carbon dioxide

   ii) fewer habitats

   iii) less food

   iv) loss of soil

   c) Suggest how the impact of deforestation on biodiversity could be reduced.

> **Hint**
>
> Carry out some research to find out the sort of activities that help preserve biodiversity. You may want to find out more about seed banks and conservation zones.

20 Inheritance

» From the genome to DNA

Worked example

Which two of the following cells from a person are genetically identical?
- egg cell
- liver cell
- red blood cell
- skin cell

The cells that contain identical genetic material are the liver cell and the skin cell.

The egg cell is haploid – having one copy of the genome instead of two. The red blood cell has lost all of its genetic material to make more room to carry oxygen.

Know

1 The diagram shows a short section of a DNA molecule:

a) What is the name given to the shape of this molecule?

b) Who were the scientists who were awarded a Nobel prize for discovering its structure?

c) What contribution did Rosalind Franklin make to the discovery of DNA?

2 Explain how a molecule as long as DNA can fit into the nucleus of a living cell.

Apply

1 The diagram shows all of the chromosomes from the nucleus of a cell.

a) Is this a diploid or haploid cell? Explain your answer.

b) How many molecules of DNA are shown in the diagram? Explain your answer.

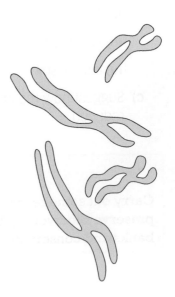

2 Copy and complete the base pairs for the section of DNA shown.

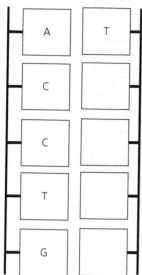

Extend

1 The table shows the diploid number of chromosomes found in the cells of different species:

Species	Number of chromosomes
Mosquito	6
Snail	24
Human	36
Fish (carp)	104

a) What are chromosomes?

b) What is meant by:

   i) diploid number

   ii) haploid number?

c) State the number of chromosomes in the gametes (sex cells) of the species listed.

d) Suggest possible explanations for the following.

   i) A human has six times as many chromosomes as a mosquito.

   ii) The carp (a fish) has many more chromosomes than a human.

> **Hint**
>
> Many species have a large number of chromosomes, and this has nothing to do with the complexity of the organism. Carry out some research to find out about polyploidy.

2 The diagram shows the process of mitosis in animal cells.

a) Describe what mitosis is used for in animals such as humans.

b) When a cell divides by mitosis it produces two new daughter cells. Use the diagram to explain why:

   i) the two cells are identical to each other

   ii) the two cells are both diploid.

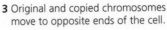

1 Chromosomes make copies of themselves and nucleus disappears.

2 Copied chromosomes line up.

3 Original and copied chromosomes move to opposite ends of the cell.

4 Cell divides.

5 New nuclei form in each of the two new cells.

3 The diagram shows the stages of meiosis:

a) Use the diagram and the mitosis diagram in question 2 to complete the following table showing some of the differences between mitosis and meiosis.

Feature	Mitosis	Meiosis
Number of daughter cells		
Number of chromosomes		
Variation between daughter cells		

b) Identify one similarity between mitosis and meiosis.

c) Meiosis is used to produce the sex cells or gametes in both plants and animals. Explain how each of the features shown in the table above is useful in the production of gametes.

d) What would happen if the chromosomes did not separate completely during meiosis?

1 Chromosomes make copies of themselves and nucleus disappears.

2 Chromosome pairs line up, and swap pieces of information (DNA crossover).

3 Cell divides.

4 Chromosomes line up.

5 Original and copied chromosomes move to opposite ends of the cell. Cell divides for a second time.

6 Four new nuclei form.

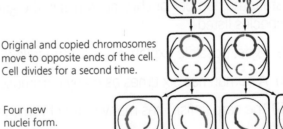

Hint

Be very careful when spelling 'mitosis' and 'meiosis'. The two processes are easily mixed up and the two words look quite similar.

» Monohybrid inheritance

Worked example

Complete the following by filling in the missing words:

A section of DNA that gives us a particular characteristic is called a _____.
We have two copies of each, known as _____, one from our _____ and
one from our _____. Cells with two copies are known as _____ and
include most of our body cells. Cells with just one copy are called _____
and include the egg and _____ cells.

A section of DNA that gives us a particular characteristic is called a
gene. We have two copies of each, known as alleles, one from our mother
and one from our father. Cells with two copies are known as diploid
and include most of our body cells. Cells with just one copy are called
haploid and include the egg and sperm cells.

Know

1 Copy and complete the table by adding either the characteristic or
phenotypes.

Characteristic	Phenotypes
Eye colour	
Blood group	A, B, AB, O
Hair colour	
Earlobe shape	
	Purple pea flowers; white pea flowers

2 A plant with red berries was crossed with a plant with yellow berries. All
of the new plants had red berries. Work out the following:

 a) the dominant allele

 b) the recessive allele

 c) the phenotype of the new plants

 d) the genotype of the new plants.

Apply

1 A male, XY, and a female, XX, have a baby. Copy and fill in the Punnett
square to show the chance of the baby being either male or female.

Hint

Remember
that humans
have 23 pairs of
chromosomes, with
22 pairs looking
the same. One pair
can be different: X
and X in females; X
and Y in males.

Extend

1 The sex of an individual is determined by two chromosomes – X and Y.

a) Give the genotype of a person who is:

i) male

ii) female.

b) Explain how the inheritance of the X and Y chromosomes determines the sex of an individual.

c) The table below shows the ratio of the number of males and females at different ages:

Age	Male	Female
1 year old	1.00	1.00
70 years old	0.67	1.00

i) Explain why the ratio of males to females at 1 year is equal.

ii) Suggest why there are fewer males than females at age 70.

» Mutations and their effects

Worked example

Mutations can occur in normal body cells as well as in the sex cells, eggs or sperm. Suggest why a mutation in a body cell is unlikely to lead to an evolutionary change in that species.

A mutation in a body cell will only affect the individual organism. The effects may be serious for that individual, but the mutation will not be passed on to future generations.

A mutation in a sex cell has a chance of being passed on to the next generation. Any effect will be inherited and will affect all future generations. Depending on the mutation, this may have an effect on the evolution of that species.

Know

1 Antibiotics are used to treat some infections by killing bacteria. Bacteria may show mutations that make them resistant to the antibiotics.

a) What is meant by a mutation?

b) Explain how the mutation could be an evolutionary advantage to the bacteria.

c) Suggest why the over-use of antibiotics is not encouraged.

2 Distinguish between the following pairs of terms:

a) mutagen and carcinogen

b) malignant tumours and benign tumours

c) genetic disorders and communicable diseases.

Apply

1 Cystic fibrosis is a genetic disease in which a large amount of sticky mucus is produced. Explain the consequences of excessive mucus production in the following parts of the body:

    a) lungs

    b) reproductive system

    c) digestive system.

2 Sickle-cell anaemia is a genetic condition in which some of the blood cells change shape and are destroyed by the body.

A male and female carrier of sickle-cell anaemia plan to have a baby. Copy the Punnett square below and use it to explain the chances of the baby suffering from the condition.

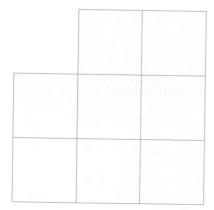

Extend

1 Mutations are permanent changes to the DNA of an organism.

    a) Suggest an example of a mutation that:

        i) is a benefit to an organism

        ii) is a disadvantage to the organism

        iii) has no effect on the organism.

    b) In a condition known as sickle-cell anaemia (SCA), the gene for haemoglobin suffers a mutation so that the haemoglobin is less effective. If a person is heterozygous for SCA then they are partially protected against malaria, a life-threatening disease.

        If two heterozygous parents for SCA have a child, show the chances of that child not suffering from SCA.

    c) How would a person with less effective haemoglobin be affected?

    d) Explain why more people have SCA around the world than would normally be expected.

Index